AI Unleashed

Industry Use Cases and Innovations

Artificial Intelligence in Organizations: Transformative Technologies, Applications, and Impacts

Print ISSN: 2979-0700
Online ISSN: 2979-0719

Series Editor: Tereza Raquel Merlo *(University of North Texas, USA)*

AI-adoption is becoming increasingly widespread. It is a transformative force reshaping countless fields and industries, revolutionizing the way organizations create and capture value. In this new world, a clear understanding of AI's implications has become indispensable for both business leaders and academic researchers. This series was developed to serve as a bridge between theory and practice, offering a holistic view of AI in business from foundational knowledge to advanced applications and future directions.

Designed to guide executives, managers, and professionals, the series explores how organizations can strategically leverage AI for innovation, sustainable growth, and competitive advantage. At the same time, the books of this series provide scholars and students with a deeper understanding of current trends, opportunities, and challenges, including ethical considerations, regulatory shifts, and workforce transformation. By integrating academic rigor with practical relevance, this series equips its readers with the knowledge, strategies, and foresight necessary to successfully navigate and lead in an AI-powered future.

Published

Vol. 3 *AI Unleashed: Industry Use Cases and Innovations*
by Marc Helmold, Ayse Kucuk Yilmaz, Tracy Dathe and Florian Hummel

Vol. 2 *Navigating Inequities and Social Justice in the Age of Artificial Intelligence*
edited by Kendra Albright, Tereza Raquel Merlo and Naresh Kumar Agarwal

Vol. 1 *AI-Driven Revolution: Transforming the Business Landscape*
edited by Tereza Raquel Merlo

Volume 3

Artificial Intelligence in Organizations: Transformative Technologies, Applications, and Impacts

AI Unleashed

Industry Use Cases and Innovations

Marc Helmold

IU International University of Applied Sciences, Berlin, Germany

Ayse Kucuk Yilmaz

Eskisehir Technical University, Türkiye

Tracy Dathe

Macromedia University of Applied Sciences, Berlin, Germany

Florian Hummel

IU International University of Applied Sciences, Berlin, Germany

World Scientific

NEW JERSEY • LONDON • SINGAPORE • BEIJING • SHANGHAI • HONG KONG • TAIPEI • CHENNAI • TOKYO

Published by

World Scientific Publishing Europe Ltd.
57 Shelton Street, Covent Garden, London WC2H 9HE
Head office: 5 Toh Tuck Link, Singapore 596224
USA office: 27 Warren Street, Suite 401-402, Hackensack, NJ 07601

Library of Congress Cataloging-in-Publication Data
Names: Helmold, Marc author | Küçük Yilmaz, Ayşe author |
Dathe, Tracy author | Hummel, Florian author
Title: AI Unleashed : Industry Use Cases and Innovations / Marc Helmold (IU International University of Applied Sciences, Berlin, Germany), Ayse Kucuk Yilmaz (Eskisehir Technical University, Türkiye), Tracy Dathe (Macromedia University of Applied Sciences, Berlin, Germany), Florian Hummel (IU International University of Applied Sciences, Berlin, Germany).
Other titles: Artificial Intelligence in Organizations: Transformative Technologies, Applications, and Impacts Volume 3
Description: Hackensack, NJ : World Scientific, [2026] | Series: Artificial Intelligence in Organizations: Transformative Technologies, Applications, and Impacts ; Volume 3 | Includes bibliographical references.
Identifiers: LCCN 2025033048 | ISBN 9781800618046 hardcover |
ISBN 9781800618053 ebook | ISBN 9781800618060 ebook other
Subjects: LCSH: Artificial intelligence--Business applications--Case studies | LCGFT: Case studies
Classification: LCC HD45 .H38 2026
LC record available at https://lccn.loc.gov/2025033048

British Library Cataloguing-in-Publication Data
A catalogue record for this book is available from the British Library.

For any available supplementary material, please visit
https://www.worldscientific.com/worldscibooks/10.1142/Q0530#t=suppl

Desk Editors: Nambirajan Karuppiah/Gabriel Rawlinson

Typeset by Stallion Press
Email: enquiries@stallionpress.com

About the Authors

Marc Helmold (M.B.A.) is a full-time professor at IU International University of Applied Sciences in Berlin. He teaches bachelor's, master's, and M.B.A. courses in business management, operations management, lean management, quality management, and leadership. He has served in several executive leadership positions at OEMs in the automotive, railway, and aviation industries. In addition to his professorial role at IU, which he has held since 2016, he conducts lean, quality, and leadership workshops in his own consultancy. Prof. Helmold has contributed to a wide range of publications.

Ayse Kucuk Yilmaz is a full-time professor in the Faculty of Aeronautics and Astronautics at Eskisehir Technical University, Türkiye. She has over 25 years of experience as an academician, researcher, and administrator, including six years as an international airport business manager and eight years in various managerial roles. She has published over 40 books, more than 70 peer-reviewed journal articles, and over 50 conference papers. She also leads and manages Erasmus+ EU projects. Her fields of expertise include enterprise risk management, risk analysis, leadership, lean management, airport management, supply chain management, digitalisation, multi-criteria decision-making, digital transformation, and digital technologies.

Tracy Dathe is a full-time business management professor at Macromedia University of Applied Sciences, Berlin, Germany. She has a rich background in finance, previously holding the position of chief financial officer (CFO) for a multinational medium-sized automotive supplier. In this capacity, she oversaw commercial management across various legal entities in Germany, China, France, Italy, Sweden, the Czech Republic, Türkiye, and the U.S.

Florian Hummel has been a professor of strategic management and vice rector for international affairs at IU International University of Applied Sciences, Berlin, since 2017. He holds a bachelor's degree from the University of Brighton and a master's degree from Cranfield University, UK. He wrote his doctoral thesis at the University of Sheffield, UK, on "An Analysis of the Impact of the Bologna Declaration on Higher and Vocational Education". Prof. Hummel began his career in a large U.S. company before moving to the higher education sector. In the past 20 years, he has worked at various universities in various roles – as a professor, dean, vice president, and managing director. He has been an external assessor at several British universities. His particular area of interest is international university cooperation and accreditation, which includes the development of double-degree programmes and articulation agreements.

Contents

List of Figures

List of Tables

Chapter 1

Artificial Intelligence (AI), Machine Learning (ML), and Deep Learning (DL)

1.1 AI, ML, and DL: Scope and Definition

In today's digital era, terms such as artificial intelligence (AI), machine learning (ML) and deep learning (DL) are ubiquitous. But what do they actually mean, and how do they differ from one another? In this short introduction, we'll take an in-depth look at the classification of these methods and their respective applications. Figure 1.1 illustrates the three elements: AI, ML, and DL.

The development of AI, ML, and DL is progressing steadily and is already having a noticeable impact on various industries. Future advances could lead to even more powerful and intelligent systems that solve complex problems and possess human-like capabilities.

Overall, the classification of AI, ML, and DL is crucial to fully exploiting their potential and understanding their applications. By better understanding these technologies, companies can develop innovative solutions and gain a competitive advantage.

The global AI market size is calculated at US$757.58 billion in 2025 and is forecast to reach around US$3,680.47 billion by 2034, accelerating at a CAGR of 19.20 percent from 2025 to 2034. The North America AI market size surpassed US$235.63 billion in 2024 and is expanding at a CAGR of 19.22 percent during the forecast period. The market sizing and forecasts are revenue-based (million/billion US$), with 2024 as the base year. Table 1.1 shows the projected market size of AI from 2024 to 2034 (Precedence, 2025). North America has garnered the largest market

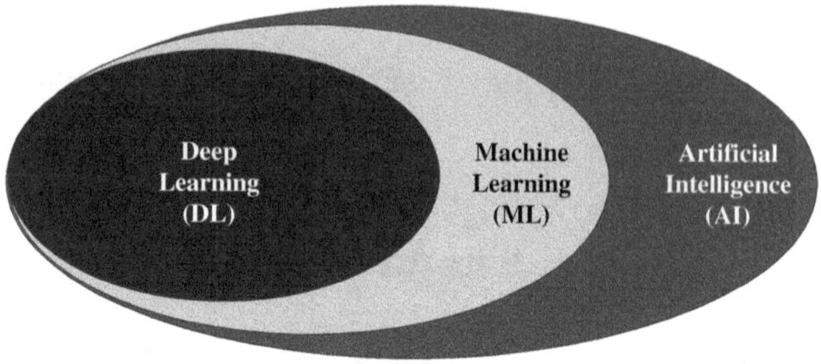

Figure 1.1. Artificial intelligence (AI), machine learning (ML), and deep learning (DL).

Table 1.1. AI market size (in US$ billion): 2024–2034.

2024	638.32
2025	757.58
2026	900.00
2027	1,070.10
2028	1,273.42
2029	1,516.64
2030	1,807.84
2031	2,196.76
2032	2,575.16
2033	3,077.32
2034	3,680.47

share in 2024. The increased demand for automated and technologically advanced hardware and software products across various end-use verticals, coupled with favourable government policies encouraging the adoption of AI by industries in North America, has significantly contributed to the growth of the AI market. In 2019, the American president launched an initiative to promote the U.S. as a leader in AI technology. This initiative focused on the adoption of AI-based systems by providing guidelines for the real-life application of AI technology in various industries and sectors. North America is home to leading tech giants such as Meta, Amazon, Google, IBM, Microsoft, and Apple, which have made significant contributions to the development of the AI market in the region. In the realm of AI markets, North America emerges as the dominant player, a position bolstered by supportive government initiatives aimed at fostering AI adoption across a multitude of industries. Significant investments in AI research and development have led to the establishment of specialist research institutes and centres that focus on advancing AI technologies. These governmental efforts not only emphasise safety and innovation but also extend into critical sectors such as healthcare, where AI is revolutionising diagnostics and treatment methodologies. Companies and research institutions in the U.S. are leading the charge in crafting groundbreaking robotic technologies that harness the power of AI to perform a diverse array of tasks, from industrial manufacturing to everyday household chores. The relentless pursuit of AI innovation, combined with public and private sector collaboration, positions North America at the forefront of the global AI landscape, driving advancements that could reshape economies and societies alike (Precedence, 2025).

1.1.1 Artificial intelligence

AI is a discipline within the branch of computer science. It describes the ability of machines to independently perform tasks that normally require human intelligence, such as logical reasoning, problem-solving, and learning from experience. Subfields of AI include speech recognition, image processing, and decision-making. AI systems are developed either on rule-based systems or through machine learning.

1.1.2 Machine learning

ML is a subfield of AI that focuses on developing algorithms that can learn from data without being explicitly programmed. By generating knowledge from experience, artificial systems learn from examples and can generalise after the learning phase is complete. This means that the system doesn't simply memorise examples but recognises patterns and regularities in the training data.

Through transfer learning, the system can also evaluate unknown data or fail to learn from them due to overfitting the model to the training data. ML models are used in various applications such as spam detection, facial recognition, medical diagnosis, and financial forecasting.

1.1.3 Deep learning

DL is an advanced form of ML based on artificial neural networks. The networks consist of multiple layers of neurons that process data through complex transformations.

DL is particularly effective at processing unstructured data such as images, text, and speech. Applications of DL include image recognition, speech translation, and autonomous vehicles.

1.2 Differences in Methods, Tools, Areas of Application, and Use Cases

The primary difference between AI, ML, and DL lies in their complexity and ability to recognise patterns in data. While AI is the umbrella term for machine intelligence, ML is a technique that enables machines to learn from experience, and DL is a specific machine learning method based on deep neural networks. Table 1.2 shows the differences among AI, ML, and DL.

Table 1.2. AI, ML, and DL.

	Artificial Intelligence (AI)	Machine Learning (ML)	Deep Learning (DL)
Methods	Algorithms, decision trees	Linear regression, random forest, gradient boosting, logical regression, k-nearest neighbour, support vector machines, k-means clustering	Artificial neural networks, convolutional neural networks, neural networks, auto encoders
Applications	Chatbots, expert systems, automation	Recommendation systems, diagnostic analysis, fraud detection	Speech and image recognition, autonomous driving
Areas of Usage	Systems of problem-solving and decision-making; decisions based on structured, rule-based and repetitive algorithms	Regression analysis, classification methods, clustering based on structural data	Processing of unstructured data such as images, text passages, or speech
Examples of Tools and Technologies	Python, R, Java, C++	SciKit-learn, XGBoost, Caret	Keras, Tensor Flow, PyTorch

1.3 AI Is Reshaping Industries, Organisations, and Employees

AI adoption in the workplace enhances processes, productivity, and incomes while simultaneously replacing and creating jobs. The reskilling and continuous professional development of staff and employees is an essential requirement for enterprises and organisations in order to achieve a competitive advantage. With regard to the future of work. AI is redesigning the workplace, transforming job roles and redefining the skills required for the future. It is therefore important that enterprises understand the necessity for a strategic transformation. AI transformation is a strategic initiative whereby a business adopts and integrates AI into its operations, products, and services to drive innovation, efficiency, and growth.

1.4 Understanding AI and Digitalisation as a Competitive Advantage

AI has become the catalyst for innovation that gives enterprises a competitive edge. It is important that AI initiatives must align closely with business goals and customer needs (Helmold *et al.*, 2024). Companies seeking maximum benefit from AI must be adaptable and open to continuous AI innovation.

Enterprises must consider creating a centre of excellence (CoE) to explore new AI applications across the value chain, including supply chain partners. It is crucial to encourage managers and staff to embrace AI and suggest new ways to improve operations and to encourage company stakeholders to collaborate on new AI initiatives. AI must be an ongoing part of the strategic roadmap in a hostile and competitive environment. The major areas of innovation in AI include the following:

- Generative AI
- Intelligent automation
- Cybersecurity
- Data analytics

1.5 Significant Innovations in AI and Digitalisation

1.5.1 Generative AI will dominate organisations and industries

Generative AI will continue its development, propelled by substantial investments in the field. These investments will drive continued progress in generative AI, fostering the emergence of new tools for text, video, image, and audio generation. Generative AI will be utilised in leading industries such as energy, healthcare, defence, and automotive (Sethi *et al.*, 2025). The size of the generative AI market is projected to reach a total volume of US$66.89 billion in 2025. The market size is expected to show an annual growth rate (CAGR 2025–2031) of 36.99 percent, resulting in a market volume of US$442.07 billion by 2031.

In global comparison, the largest market size will be the U.S. (US$21.65 billion in 2025) (Statista, 2025).

1.5.2 Cognitive systems and machines

Cognitive systems are able to independently develop solutions for human tasks. Cognitive systems and machines are technical systems that collect digital information from sensor data and networks and derive conclusions, decisions, and measures based on learning algorithms. These elements are continuously checked and optimised in cognitive systems (Portmann and D'Onofrio, 2020). Cognitions are procedures or processes with which the individual recognises himself and his environment, i.e., they are processes of mental information processing. They serve for mental control and voluntary control of behaviour. Cognitive technologies mimic the functions of the human brain. These include, for example, natural language processing, data mining, and pattern recognition.

1.5.3 Machines will learn

Machines will become smarter and autonomously learn by utilising algorithms. This is the field of ML and its related algorithms (Helmold, 2024). ML algorithms are split into two main categories based on how they interact with data: supervised and unsupervised. ML is a sub-discipline of digitalisation and AI. The focus is on training computers, machines, or devices to learn from data and experience and continually improve, rather than being explicitly programmed to do so. An artificial system learns from examples and can generalise them after the learning phase has ended (Nitschke, 2021). ML trains algorithms to find patterns and correlations in large data sets and make the best decisions and predictions based on this analysis. ML applications improve with use and become more accurate the more data are available.

1.5.4 Machines will make decisions

Automated decision-making systems (ADMSs) are technical systems that support or completely autonomously make complex decisions. Although the consequences for those affected can be significant, they are often unaware that an ADMS was used and cannot understand or challenge the decision-making process. The digital society is therefore calling for the regulation of ADMSs, ensuring that the use of these systems is transparent and comprehensible and that ADMSs may not be used if the associated risks are unacceptably high. Social media algorithms determine what information we see and what we don't. Application portals automatically create personality profiles for candidates. The predictions of a predictive policing system guide the deployment of police patrols. These and other examples illustrate the growing importance of ADMSs in our lives. But society as a whole is also influenced by them, for example when ADMSs disadvantage or favour certain groups of people due to systematic biases. A critical reflection on whether and how ADMSs should be used is therefore needed (Helmold, 2024).

1.5.5 Neural networks

A new spirit of optimism has arisen due to the great successes of neural networks in recent years, using large data collections as the key to success. A neural network recognises patterns in a similar way to the human brain, as it has many neurons connected to one another. The knowledge lies in the connections (edges within the brain: synapses) that determine how strongly a signal is transmitted from one neuron to another. The neurons themselves add the incoming signals and pass them on to all outputs according to the sum (modulated by an activation function; in human neurons with an all-or-none threshold). The knowledge in neural networks or the brain is not located in a single edge or synapse but is distributed across many. The strength of the edges is learnt from examples.

1.5.6 Neural coding

Neural coding (or neural representation) is a neuroscientific field concerned with characterising the hypothesised relationship between the stimulus and the individual or ensemble neuronal responses, as well as the relationship between the electrical activity of the neurons in the ensemble. Based on the theory that sensory and other information in the brain are represented by networks of neurons, it is believed that neurons can encode both digital and analogue information. Neurons stand out among the body's cells for their ability to quickly propagate signals over long distances. To do this, they generate characteristic electrical impulses – so-called action potentials: voltage peaks that can extend across axons. Sensory neurons change their activities by firing sequences of action potentials in different temporal patterns when external sensory stimuli such as light, sound, taste, smell, and touch are present. Information about the stimulus is known to be encoded in this pattern of action potentials and transmitted within and around the brain, but this is not the only method. Specialised neurons, such as those in the retina, can transmit more information through graded potentials. This is different from action potentials because information about the strength of a stimulus directly correlates with the strength of the neuron's output. With graded potentials, the signal decays much faster, requiring shorter distances between neurons and higher neuronal density. The advantage of graded potentials is higher information rates, allowing more states to be encoded (i.e., higher fidelity) than spike neurons. The connection between stimulus and response can be examined from two opposing perspectives. Neural coding refers to the mapping from stimulus to response. The main focus is on understanding how neurons respond to a variety of stimuli and constructing models that attempt to predict responses to other stimuli. Neural decoding refers to the reverse mapping from response to stimulus, where the challenge is to reconstruct a stimulus or certain aspects of that stimulus from the spike sequences it evokes.

1.5.7 Using AI in normal life

AI encompasses all efforts aimed at making machines intelligent in business and private environments (Helmold, 2024). Intelligence is understood as the quality that enables a being to act appropriately and proactively in its environment. This includes the ability to perceive and react to sensory impressions, to absorb information, process it, and store it as knowledge, understand and produce language, solve problems, and achieve goals. Since the term was coined in 1955, a number of relatively independent sub-disciplines have emerged, such as:

- pattern recognition, which also includes speech recognition and handwriting recognition;
- knowledge modelling, including logical programming and inference engines;
- expert systems, question-answer systems, and chatbots;
- ML;
- artificial neural networks and DL;
- computer vision;
- robotics and automatic transportation;
- universal game programs.

1.5.8 Deep learning

DL is part of a larger family of ML methods based on artificial neural networks with representation learning. In DL, the adjective "deep" refers to the use of multiple layers in the network. The methods used can be either supervised, semi-supervised, or unsupervised. DL architectures such as deep neural networks, deep belief networks, deep reinforcement learning, recurrent neural networks, convolutional neural networks, and transformers have been applied in areas such as computer vision, speech recognition, natural language processing, machine translation, bioinformatics, drug design, medical image analysis, climate science, materials inspection, and board game programs, where they have achieved results comparable to or, in some cases, exceeding the performance of human experts. Artificial neural networks are a basis for modelled ML methods utilising the network of nerve cells in the brain. They consist of data nodes and weighted connections between them.

1.5.9 Black box and white box testing

Black box testing and white box testing are two AI methods that can be used to check components, code, and operability (BMWi, 2020). Due to their differences, both test methods are suitable for very different scenarios in practice. Black box models of ML are purely statistical models. White box models, on the other hand, refer to analytical and physical descriptions, the modelling of which is usually very complex. Grey box models combine both approaches to combine the respective advantages.

1.5.10 Using neuromorphic chips

Neuromorphic chips are microchips in which the architecture and properties of neurons are modelled at the hardware level. These neuron-like components simulate the brain's learning and association powers, which can accelerate particularly in recognising patterns in images or large data structures.

1.6 Seven Patterns of AI

From autonomous vehicles, predictive analytics, and facial recognition to chatbots, virtual assistants, cognitive automation, and fraud detection, the use cases for AI are diverse. However, regardless of the application of AI, there are commonalities across all of these applications. Those who have implemented hundreds or even thousands of AI projects recognise that, despite the variety of applications, AI use cases follow one or more of seven common patterns. The seven patterns are hyper-personalisation, autonomous systems, predictive analytics and decision support, conversation/human interactions, pattern and anomaly detection, recognition systems, and goal-oriented systems. Each tailored approach to AI requires its own programming and patterns, but whatever combination of trends is used, they all adhere to fairly standardised rules. These seven patterns are then applied individually or in various combinations, depending on the specific solution to which AI is being applied. Figure 1.2 shows the seven patterns of AI.

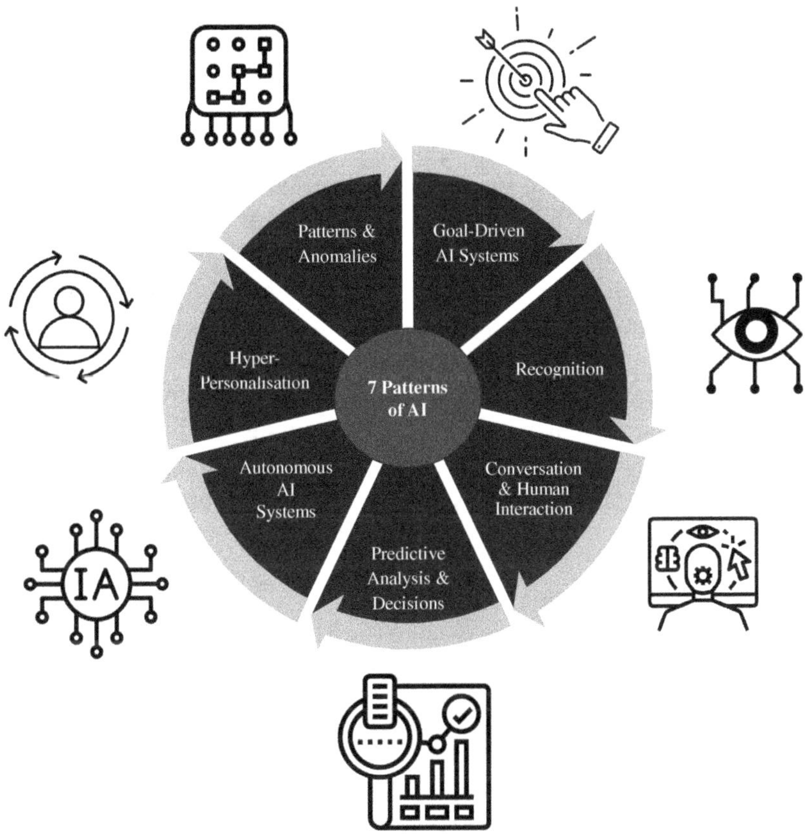

Figure 1.2. Seven patterns of AI, ML, and DL.

1.7 Trends and Outlook in AI

1.7.1 AI drives value chains and market opportunities

In the future, AI will assess and drive business opportunities throughout the value chain. The primary focus of AI usage is to streamline routine tasks and improve business efficiencies. Companies should clearly understand why they need AI and what results they want to achieve with its help.

It is crucial for each company to define its business goals and to understand AI's capabilities in this direction.

AI technologies optimise this chain by enhancing supply chain management, logistics, reducing waste, and improving quality control. AI algorithms can optimise transport routes to minimise delivery times and costs, thereby reducing spoilage and ensuring fresher produce reaches consumers promptly. AI will forecast market opportunities by using programs, tools, or predictive analyses (Helmold, 2024). In the early days of AI, projects were driven exclusively by data scientists. They had the data, the algorithms, and the freedom to look for ways to address business problems with their new tools. Sometimes they were successful, but today that dynamic has been reversed. Business leaders have learnt from successful projects and have a better understanding of what AI can do for them. That is why companies are now less willing to invest in proof-of-concept projects where the added value for the business is unclear. This means that business units are increasingly in charge when AI is introduced.

1.7.2 AI creates jobs, AI is not a job terminator

AI will help enterprises to eliminate the need for human activities that are repetitive and monotonous. In doing so, it continues a process that began with industrial society. It will not be linear or quantitative, in the sense of making jobs redundant. But rather as a constant shift from uncreative to more creative activities, from isolated to more communicative activities. This creates stress in society but is also a liberation from possibilities that were previously hidden under monotony. Many jobs involve explicit or implicit human factors, even if they involve strong routines. Nurses don't simply "care for" sick people; they have a relationship with them. Bartenders don't simply shake cocktails; they practise soul communication. Journalists do not produce information; they produce humane interpretations. AI systems can pool knowledge and thereby create space for human empathy. AI is shifting the professional spectrum towards greater complexity. Traditional professions that are in danger of becoming "Taylorised" can find their way back to their origins. Doctors can heal instead of processing. Journalists can interpret again rather than creating floods of information. Nurses can again exercise empathy instead of managing patients. Craftsmen are once again designing with their hands instead of assembling. Traders can trade again instead of calculating. At the same time, the principle of differentiation creates countless new fields of activity: mediators and moderators, connectors and curators, coaches and life companions, traffic managers and healthcare providers, mindfulness agents, and beauty designers. The number of these professions will exceed the number of activities in industrial society and will enable human creativity to be unleashed.

1.7.3 AI will trigger the next industrial revolutions

Industrial Revolution 4.0 involves the convergence of AI-based intelligence and computer technologies at one point to create a smart factory system. The system works to develop focused intelligent networks, machines, and functions like humans. AI has made significant strides in automating administrative tasks that are often time-consuming and repetitive. Through AI-driven software and applications, businesses can streamline paperwork processes, manage scheduling efficiently, and handle routine office operations with greater ease. AI has brought the Fourth Industrial Revolution to an inflection point, and manufacturers must choose a path forward: innovate, accelerate, or follow fast.

In the future, AI will trigger further industrial revolutions. AI, with its tools such as ChatGPT, will radically transform industry, value chains, and society and trigger an industrial revolution. All information-based value creation processes will radically change and reform. AI can already compose and paint, write journalistic texts, solve exams, and generate text modules containing legal arguments.

1.7.4 Computers do not yet feel but will develop biometrical interpretations

In science fiction films, intelligent and sentient computers and machines have long been a reality. AI systems and computers interact with their environment and users and react to stimuli with artificial sensory perceptions. But how realistic are these ideas? In biological life forms, the brain can combine information from different senses to create a representative image of its environment. For example, the characteristic scent, the sight of a dark brown liquid, and a gurgling noise combine to create a representation of coffee straight away. Similarly, although we can identify a kiwi just by looking at it, whether it is ripe or not is only revealed by the smell and a slight pressure on its shell. In the event that sensory perceptions, stimuli, and emotions can be explained biochemically, AI interpretation in the future is certainly conceivable. However, unlike humans, AI cannot yet feel (Gumm, 2008). Moreover, today AI is not yet able to achieve a complete understanding of emotions, intelligence, creativity, or consciousness, although many elements can be explained through biochemistry, physics, and mathematics.

1.7.5 Symbiosis of AI and humans

AI is already a large part of our daily business and social lives. Whether it's a smartphone, autonomous driving, or implants for certain bodily functions, there is artificial support for almost every area of life. In the future, humans and AI will form even more symbiotic units (Rainy, 2023). In numerous research areas such as healthcare, industry, logistics, and medicine, there will soon be even more viable ways in which people and AI become symbiotic units (Helmold, 2024).

1.7.6 Upgrade of human intelligence through AI

The use of AI will lead to an upgrade of human intelligence. The so-called cyber-physical systems are autonomous, intelligent, embedded in their environment, and can interact with each other and with the environment. While AI provides us with improved models, forecasts, and simulations, it also challenges humans to improve possible future conditions. The social media crisis shows that it is not enough to develop optimising algorithms. The increase in efficiency of human communication leads to enormous consequential damage. Human communication relies on trust, reciprocity, and patience. Oppression and mental suffering are the result when communication is designed exclusively within the categories of increasing stimuli and reactions (Zukunftsinstitut, 2023).

1.7.7 Innovation push in healthcare and sustainability through AI

The greatest spurts of innovation through AI are currently occurring in medical technology and in the area of sustainability. AI in medicine has been used successfully for years. DeepMind, for example, the subsidiary of Google's parent company Alphabet, uses a platform to help predict protein structures (Helmold, 2023). The structure of a protein determines its function in the cell. If this structure is known, drugs can be developed more precisely. What is new are AI-based processes that recreate proteins. To do this, researchers fed a speech generator with amino acids, the building blocks of proteins, and information about their function. Only a fraction of the resulting proteins were able to kill bacteria like real proteins. In this way, new medications and medical devices can be developed (Krolle, 2023). In the area of sustainability, the greatest innovation spurts come from the development of technologies that remove CO_2 from the air and process it as a raw material, for example in the production of building materials, kerosene, cosmetics, and food (Helmhold, 2021).

1.7.8 Forecasting will become accurate through AI

For a long time, forecasting was a marginal science; predictions were often more magic than science. Now, a change is taking place: more science, less fiction. Because of digitalisation, predictions are getting better and better. They are based primarily on the growing availability of data, the performance of computers, and ML. Algorithms are trained to recognise empirical connections in large amounts of data. For example, they learn to distinguish moles that are potentially cancerous from harmless freckles, achieving an accuracy that exceeds that of human experts. In addition to medicine, such algorithms are increasingly being used in other industries such as law, taxation, marketing, transport, and agriculture. A key difference between mathematical and magical methods is that the forecasts can be checked and compared in real time with real developments and competing forecasts, which in turn improves forecast quality. Algorithms do not confer clairvoyant abilities. They do not allow us to see the future, but they make facts and connections visible. Predictive analytics are task-specific prediction systems that do not focus on the big picture but rather on the known unknowns. The focus is on short-term or real-time forecasts, for example, for weather, traffic conditions, or epidemics.

1.7.9 Legal framing and discussions on AI

Currently, global AI is not directly anchored in legal systems through uniform rules and principles. Even the strongest AI could not currently register intellectual property rights because it would not be a legal entity. So far, AI has ultimately been another tool that inventors use in their work. It is undisputed that AI can improve the quality of life of many people and help overcome global challenges such as climate change or health crises. A significant amount of money is already being made available worldwide for the development of AI systems. This shows how high the economic potential of AI is. However, there are also dangers, for example, from hacker groups or cyberattacks on public and important infrastructure. The European Union (EU) AI Act is the first AI law published by a major international regulator – a regulation establishing harmonised rules. It successfully passed the European Parliament on 14 June 2023 with a clear majority. The law covers the control and monitoring of AI in the EU.

Bibliography

Fraunhofer (2023) 'Trends in Artificial Intelligence'. Available at: https://www.fraunhofer.de/content/dam/zv/en/Publications/Trends-in-artificial-intelligence.pdf (Accessed: 20 September 2023).

Gumm, M. (2008) *Mensch, Technik, Ärger? zur Beherrschbarkeit soziotechnischer Dynamik aus transdisziplinärer Sicht*. Münster: LIT Verlag.

Hamdan, A. and Harraf, A. (2024) *Business Development Via AI and Digitalization. Volume 1*. Cham: Springer.

Hattburg, A.T. (2020) *Digitalisierung als Chancengeber: Wie KI, 3D-Druck, Virtual Reality und Co. neue berufliche Perspektiven eröffnen*. Wiesbaden: Springer.

Helmold, M. (2021) *Kaizen, Lean Management und Digitalisierung: Mit den japanischen Konzepten Wettbewerbsvorteile für das Unternehmen erzielen*. Wiesbaden: Springer.

Helmold, M. (2023) *Wettbewerbsvorteile entlang der Supply Chain sichern: Best-Practice-Beispiele in Beschaffung, Produktion, Marketing und anderen Funktionen der betriebswirtschaftlichen Wertschöpfungskette*. Wiesbaden: Springer.

Helmold, M. (2024) *Erfolgreiche Transformation zum digitalen Champion: Wettbewerbsvorteile durch Digitalisierung und Künstliche Intelligenz*. Wiesbaden: Springer.

Krolle, H. (2023) 'Sieben Thesen über die Zukunft Künstlicher Intelligenz', *Handelsblatt*, 21 October. Available at: https://amp2.handelsblatt.com/technik/forschung-medizin/handelsblatt-kamingespraech-sieben-thesen-ueber-die-zukunft-kuenstlicher-intelligenz-/29005442.html (Accessed: 21 October 2023).

Nitschke, M. (2021) *Künstliche Intelligenz und Maschinelles Lernen: Vergleich der Deep Learning Frameworks TensorFlow, Keras, PyTorch, CNTK und MXNeT*. Stuttgart: Grin Verlag.

Portmann, E. and D'Onofrio, S. (2020) *Cognitive Computing: Theorie, Technik und Praxis*. Wiesbaden: Springer.

Precedence Research (2025) 'Artificial Intelligence (AI) Market Size, Share, and Trends 2025 to 2034'. Available at: https://www.precedenceresearch.com/artificial-intelligence-market (Accessed: 15 May 2025).

Rainy, B. (2023) 'Symbiose zwischen Mensch und künstlicher Intelligenz?', *Max-Planck-Institut*, 21 October. Available at: https://www.mpg.de/19783810/symbiose-zwischen-mensch-und-kuenstlicher-intelligenz (Accessed: 21 October 2023).

Sethi, G. *et al.* (2025) *Revolutionizing Healthcare 5.0: The Power of Generative AI. Advancements in Patient Care Through Generative AI Algorithms*. Heidelberg: Springer.

Statista (2025) 'Generative AI – Worldwide'. Available at: https://www.statista.com/outlook/tmo/artificial-intelligence/generative-ai/worldwide (Accessed: 15 May 2025).

Vagadia, B. (2020) *Digital Disruption: Implications and Opportunities for Economies, Society, Policy Makers and Business Leaders*. Cham: Springer.

Zukunftsinstitut (2023) '6 Thesen zur Künstlichen Intelligenz'. Available at: https://www.zukunftsinstitut.de/artikel/digitalisierung/6-thesen-zur-kuenstlichen-intelligenz/ (Accessed: 21 October 2023).

Chapter 2

AI Transformation and Roadmap: The Competitive Advantage

2.1 Digital Transformation and AI: Scope and Definition

Digital transformation is crucial for businesses to thrive in today's competitive market. This process involves leveraging technology to improve efficiency, enhance customer experiences, and stay ahead of the competition. By embracing digital transformation, companies can streamline operations, innovate, and ultimately achieve greater success (Helmold *et al.*, 2024). Transformation is a natural process that occurs in all living things, and this is essential for survival and progress. Transformation in individuals is often associated with personal growth or self-improvement. Personal transformation can happen through self-reflection, learning new skills, or adapting to new environments (Wolan, 2024).

The advantages of digitalisation are shown in Table 2.1.

Digital transformation focuses on more agile, innovative, and customer-centric organisations using artificial intelligence (AI). AI involves the creation and use of algorithms and systems that perform tasks typically requiring human intelligence, such as learning, reasoning, problem-solving, and understanding natural language. AI is a catalyst for the successful transformation of enterprises. Digital transformation, including AI adoption, has become one of the highest priorities for public organisations (Hamdan and Harraf, 2024). For instance, enterprises or

Table 2.1. Advantages of digitalisation.

Competitive Advantage	Digital transformation allows businesses to innovate and adapt quickly, giving them a competitive edge over rivals.
Improved Efficiency	By automating processes and leveraging technology, companies can increase productivity and reduce costs.
Enhanced Customer Experience	Digital transformation enables businesses to personalise customer interactions, improve service, and build stronger relationships.
Innovation and Growth	Embracing new technologies and adapting to the digital landscape open doors to new opportunities and increase growth potential.
Agility and Flexibility	Digital transformation promotes agility and flexibility, allowing businesses to respond quickly to changing market conditions.
Data-Driven Decisions	Digital transformation promotes agility and flexibility, allowing businesses to respond quickly to changing market conditions.
Attract and Retain Talent	A digitally transformed organisation can attract and retain top talent by offering a modern and innovative work environment.

governments are increasingly trying to digitalise their services through technology, thanks to growing public expectations.

2.2 The Roadmap for Digitalisation and AI

Digitalisation and AI are keywords and popular terms used in all areas of society and include future-oriented innovations and technologies. The digital transformation and the use of human-like (humanoid) machines are among the most important social and economic developments of our times (Schallmo, 2023). Many basic questions are associated with these: What exactly is behind terms such as *digital change* and *digital transformation of AI*?

Digitalisation refers to the use of data and algorithmic systems for new or improved processes, products, and business models. In this context, digitalisation includes the conversion of analog, continuously displayable values, or the recording of information about physical objects in formats that are suitable for processing or storage in digital technical systems (Sendler, 2020).

AI uses computers and machines to mimic the problem-solving and decision-making abilities of the human mind. Intelligence is understood as the property that enables a being to act appropriately and proactively in its environment. This includes the ability to perceive and react to sensory impressions, to absorb information, process and store it as knowledge, understand and produce language, solve problems, and achieve goals.

Digitalisation and AI are revolutionising value chains, businesses, and human life, thereby unlocking a whole new world of innovation. As AI becomes more prevalent, it will affect nearly every aspect of society, from our professional to personal lives.

Figure 2.1 shows the major elements and key aspects of AI and AI applications. It is evident that AI consists of a variety of digital and intelligent applications and elements. Robots will increase economic growth and productivity, and create new career opportunities for many people worldwide. Robots are more accurate and efficient compared to humans, minimising medical and industrial errors. Robots are cost-effective, specifically as a long-term investment for businesses. In addition, robots can produce high-quality products. Industrial robots are used in manufacturing, logistics, agriculture, construction, and other industries. Robots are used for welding, painting, assembly, and testing in the automotive industry. Robots can perform these tasks more accurately and consistently

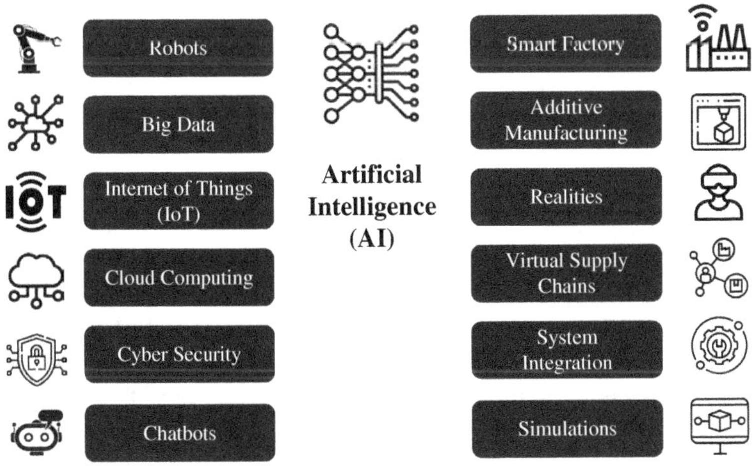

Figure 2.1. What is AI?

than humans. Automotive companies also use industrial robots to make them more competitive by increasing efficiency and lowering manufacturing costs. In addition to using industrial robots in their own production facilities, many automotive companies buy robotic automation solutions from suppliers as part of their outsourcing programs or business models that allow customers to lease or rent automated equipment rather than purchase it outright (known as "factory-less manufacturing"). Robots will also be used in areas such as agriculture, healthcare, military, and education in the future to assist humans in their daily lives (Helmold, 2024). Value chains will be generating a large quantity of data. Big Data primarily refers to datasets that are too large or complex to be dealt with by traditional data-processing application software. Data with many entries offer greater statistical power, while data with higher complexity may lead to a higher false discovery rate. This Big Data will be driven by the Internet of Things (IoT). The IoT describes devices with sensors, processing ability, software, and other technologies that connect and exchange data with other devices and systems over the Internet or other communications networks. The Internet of Things encompasses electronics, communication, and computer science engineering. Cloud systems and cloud computing will integrate internal and external supply chain partners, using a common platform of data. Products will be produced in smart factories, which use digital twins to simulate, plan, and optimise processes. Technologies such as additive manufacturing, reality simulations, virtual supply chains, system integration, and ChatGPT will support the entire value chain.

2.3 Digital Maturity Levels

A number of market forces are now driving the need to become digital. The vast majority of companies are just embarking on complex digital transformation journeys, encompassing all aspects of their business to redefine how they operate. Digital maturity refers to the ability of a company to quickly respond to the developments and shifting trends of technology (Dieffenbacher, 2022). The focus is on creating value through these new capabilities and consumer interests fast enough to have a competitive advantage over rivals. Digital maturity requires investment in the tools and human resources that can best leverage new technology.

2.3.1 Digital champions

Digital champions are role models and pioneers for digital transformation within value chains, society, and organisations. These are companies offering innovative, digital products and services that generate real added value for their employees and customers (Figure 2.2). Digital champions manage to build particularly strong digital customer proximity with their transformation towards digitalisation and an online presence on websites, social media profiles, and apps.

Figure 2.2 shows the respective maturity levels and roadmap from the digital laggard to the digital leader.

2.3.2 Digital laggards

Laggards take a narrower approach and apply new technology to improve what they are already doing. While they certainly see a benefit, they do little to prepare themselves for the growing tide of nimble startups and dynamic competitors beginning to emerge (Vagadia, 2020).

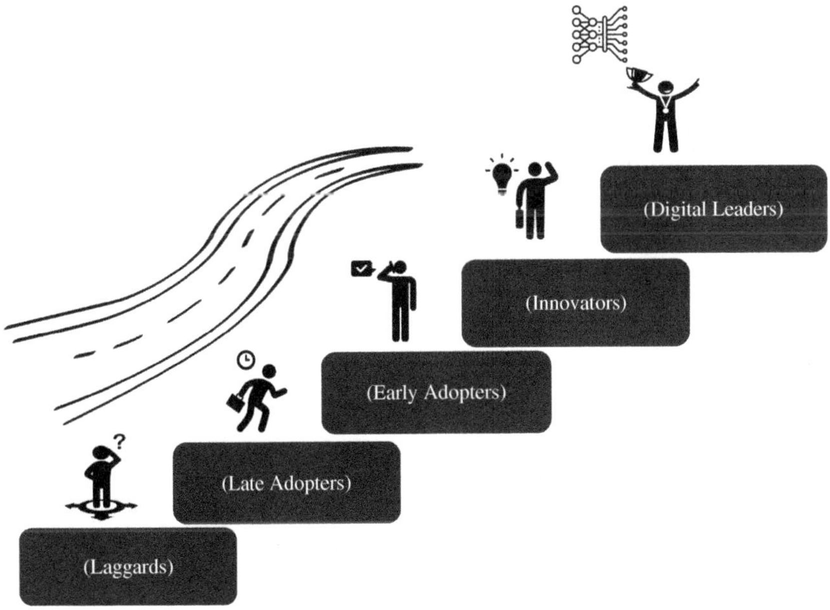

Figure 2.2. Roadmap from digital laggard to digital leader.

Digital laggards are companies that have not yet understood the need for digital transformation. Laggards have a significant gap compared to the pioneers of digital transformation and can be seen as last or latecomers, lagging far behind the other companies when it comes to digitalisation and AI. Laggards often fall by the wayside in competition due to their hesitation (Helmold, 2023).

2.3.3 Late adopters

Companies in Germany primarily recognise opportunities in the use of AI and see more advantages in using the technology compared to the previous year. At the same time, the proportion of companies that use AI is only increasing very slowly (Streim, 2023). On the one hand, the late adopters or passive implementers, who have no independent drive to innovate but only react to changing customer requirements, are at the bottom of the pack. Late adopters are hesitant and passive about innovation and the introduction of AI. Many companies wait to introduce necessary innovations because they are just waiting for them to work. The adoption of digital innovations along the value chain is also a social decision-making process, whereby if one user talks to another potential user about an innovation that works for the first user, the second user is more likely to adopt that innovation.

2.3.4 Early adopters

Unlike late adopters, early adopters are not laggards in the introduction and implementation of digitalisation measures, but rather companies that partially introduce innovations after carefully weighing up all the advantages and disadvantages.

2.3.5 Innovators

The innovators are pioneers of their time and belong to the group of adopters who are the first to buy a new process or product, thus initiating the spread of innovations. Innovators have a high willingness to take risks. Digital innovations are innovations that result from the innovative use of digital technologies. A central feature of digital innovations is that they are formed by combining multiple innovative technical solutions.

2.3.6 Digital leaders

The leading drivers and impulse generators in digitalisation include the following: digital leaders, professional digital champions, embedded digital champions, and volunteer digital champions. At the forefront of innovation dynamics are the digital leaders who act as technology leaders, constantly pushing technological boundaries. They do this with the help of strong basic research and numerous of their own inventions in the area of digitalisation along the value chain. Naturally, they have a strong scientific focus in their work. They take care of the strategic implementation of all the activities. As digital leaders, they know how important digital integration and transformation are for the success of their business model. If a permanent, systematic change needs to be implemented in the company, they know the way. This is exactly what makes them indispensable as pioneers. Professional digital champions actively drive changes and digitalisation activities in companies. They are the driving force in the company. Digital leaders must carefully select professionals and deploy them in a targeted manner so that they can inspire and train their colleagues. Embedded digital champions work in a specific function and specialist departments such as purchasing, human resources, marketing, sales, and production. They integrate their commitment as digital champions into their respective positions. Without it being explicitly stated in their job title, they are champions because they regularly lend a helping hand to colleagues. If there are topics related to digitalisation or tools that require more intensive training, volunteer digital champions come into play. They also take unpaid time and support people or organisations wherever necessary.

2.4 Strategic Phases to Become a Champion

Value chains are constantly changing. Due to the growing internationalisation of businesses and intensification of partnerships, processes are becoming increasingly complex, and demands for efficiency and cooperation are increasing. There is also a growing need to evaluate information and data, and create the greatest possible transparency about markets and customers. For some time now, companies have been increasingly looking for new business areas and potential. This requires digital strategies and innovative IT solutions to achieve the goals of the business (Engelen and Schneider, 2021). This increases the pressure on management to develop

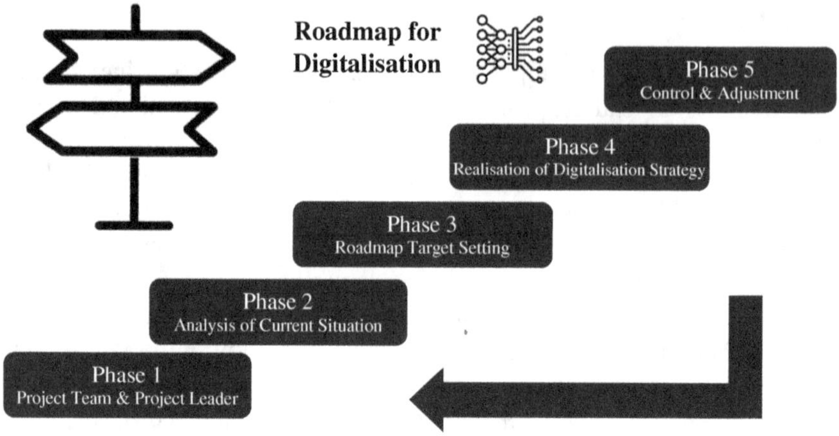

Figure 2.3. Strategic phases for digitalisation and AI.

a digital roadmap for their company (Machado and Davim, 2023). The aim of a business-oriented digitalisation roadmap is to optimally align all value creation activities of internal and external stakeholders with the company's goals and business processes (Figure 2.3). In the first phase, project managers and project teams are selected. External consultants and experts can also be involved for this purpose. In the second phase, an analysis of the current situation is conducted. In the third phase, the roadmap is delineated, and measurable, realistic, and implementable goals are defined. In the penultimate and fourth phase, the strategy is implemented. Finally, in the fifth phase, the strategy is evaluated and verified for successful and sustainable implementation.

2.5 Dealing with Digital Resistance

Resistance to digitalisation can manifest in various forms, from passive scepticism to outright refusal. Some employees may fear that new technologies will replace their roles or disrupt established workflows. Others may feel overwhelmed by the prospect of learning new systems or processes. Additionally, cultural factors within an organisation, such as established hierarchies or resistance to change, can worsen willingness to support digitalisation efforts. Resistance will also be more heightened in organisations where change fatigue is present, particularly when change efforts have been poorly managed in the past, leading to a lack of confidence in delivery. Resistance often arises when employees feel overwhelmed by the new technologies or do not understand why changes are necessary. The unpredictability and unknown of the process are the main causes of worries about digital transformation. Regular, open communication in the workplace is the best ally in overcoming fear. Enterprises must therefore have appropriate strategies, as depicted in Figure 2.3.

2.5.1 First strategy: Analyse internal stakeholder groups

Every digitalisation project has different interest groups and stakeholders (Helmold, 2022). Stakeholders (also called project participants) are people, groups of people, associations, or organisations, who are affected by the project or its results or can influence the project. Internal stakeholders, in particular, are often forgotten or underestimated; the focus is usually on the customer or the public. Experienced project managers, therefore, always analyse which groups in the company are affected by their project and then try to gain clarity about the individual groups:

- Which stakeholders are affected?
- How are the stakeholders affected?
- What are the interests of the individual groups?
- What behavior can be expected from the individual groups?
- What pattern of action are they likely to follow?

2.5.2 Second strategy: Explain the background to the transformation and digitalisation project

Projects bring changes, and changes offer opportunities. But these opportunities alone cannot convince all stakeholders about the benefits of a transformation project. Real orientation provides answers to the question of why a change project is being carried out. Only then will it be explained how the project should run and what exactly will change. In this context, it is necessary to explain the reasons for the project. Targets, time, financial, personnel or other conditions, demarcations from other projects, and project-specific organisation should also be communicated.

2.5.3 Third strategy: Deal with opponents correctly

Almost every project has opponents, and opponents are those who oppose change. The only question is how the project manager optimally behaves towards these opponents or resistance fighters and how he approaches them. Resistance can be carried out openly or covertly. Fear or insecurity is often the main reason for resistance, e.g., fear of losing cherished habits and familiar surroundings, or the fear of suffering disadvantages in the project. The biggest mistake when dealing with opponents is withholding and concealing bad news (Helmold, 2022). Since many people have a good gut feeling for the truth, they anticipate the consequences of projects. The feeling that something is being kept secret increases mistrust and resistance. This is why project professionals provide early and open information about the dark side of the project.

2.5.4 Fourth strategy: Preserve and develop what is good

Many stakeholders are not fundamentally opposed to the innovations in a project. However, they want proven and good things to be retained. Project managers should not dismiss this interest of the "keepers". On the one hand, project managers tend to do a *tabula rasa* and unnecessarily delete tried and tested things. On the other hand, it helps stakeholders if, despite all the innovations, they continue to find what they know and value. For example, they can incorporate good things that they find into the change concepts and improve them even further by cleverly combining them with the new. Experienced project managers provide clear and emphatic information about what will not be changed by the project and will remain valuable (Helmold, 2022).

2.5.5 Fifth strategy: Awareness of the valley of tears

A problem, especially with organisational change and change management projects, is the negative aspects that arise, as innovations such as converted processes and procedures or organisational changes only really work correctly after a learning curve. These negative aspects often lead to a "valley of tears" among employees. Changes disrupt a system that has often previously oscillated well; time usually passes until the changed system has stabilised again, but then at a higher level than before. Project managers call this period of time the "valley of tears", the weeks and months in which the promised benefits do not (yet) materialise. The stakeholders must therefore be made aware of the valley of tears early on and be made to understand that worthwhile change also requires patience. Otherwise, the project manager forfeits acceptance. The "valley of tears" known as the change curve is the turning point in the change process. Those affected realise that there is no way around other than saying goodbye to old habits and adapting to the new reality.

2.5.6 Sixth strategy: Provide individual information

As important as the background and procedures are in a project, the stakeholders want to know above all how the project will affect their everyday work, specifically and individually. Project managers, therefore, develop an individual information strategy for each stakeholder group. A four-step approach is recommended for individual information (Helmold, 2022): (1) Explain to each group what is valuable from their area and what will be preserved. (2) What will each group lose? What will the project change for her, and what will she have to do without in the future? (3) What does the individual gain from the change? (4) What opportunities and advantages arise for the whole, e.g., for the company or the client? "Some project managers constantly present the overall benefits to the company. Strategically, it is cleverer to talk about it at the end and first to understand and meet the needs of the individual" (Helmold, 2022).

Bibliography

Dathe, T. *et al.* (2022) *Corporate Social Responsibility (CSR), Sustainability and Environmental Social Governance (ESG). Approaches to Ethical Management.* Cham: Springer.

Dieffenbacher, S. (2022) 'What is Digital Maturity, How to Measure, Tools & Models', *Digital Leadership*, 20 March. Available at: https://digitalleadership.com/blog/digital-maturity/#What_is_Digital_Maturity (Accessed: 20 March 2025).

Engelen, A. and Schneider, O. (2021) *Die Strategien digitaler Champions.* Wiesbaden: Springer.

Forster, N. (2019) *Hidden Digital Champions: Wie sich KMUs und das Handwerk für die Zukunft rüsten (Fit for Future).* Wiesbaden: Springer.

Hamdan, A. and Harraf, A. (2024) *Business Development Via AI and Digitalization. Volume 1.* Cham: Springer.

Helmold, M. (2022) *Leadership: Agile, virtuelle und globale Führungskonzepte in Zeiten von neuen Megatrends*. Wiesbaden: Springer.

Helmold, M. (2023) *Wettbewerbsvorteile entlang der Supply Chain sichern: Best-Practice-Beispiele in Beschaffung, Produktion, Marketing und anderen Funktionen der betriebswirtschaftlichen Wertschöpfungskette*. Wiesbaden: Springer.

Helmold, M. (2024) *Erfolgreiche Transformation zum digitalen Champion: Wettbewerbsvorteile durch Digitalisierung und Künstliche Intelligenz*. Wiesbaden: Springer.

Helmold, M. *et al.* (2022) *Lean Management, Kaizen, Kata and Keiretsu: Best-Practice Examples and Industry Insights from Japanese Concepts*. Wiesbaden: Springer.

Helmold, M. *et al.* (2024) *The Digital Champion: Best Practices and Insights for the Successful Transformation of Enterprises and Companies*. Singapore: World Scientific.

Hofert, S. (2022) *Agiler führen: Einfache Maßnahmen für bessere Teamarbeit, mehr Leistung und höhere Kreativität*. Wiesbaden: Springer.

Knuppertz, T. and Ahlrichs, F. (2022) *Prozessmanagement und Agilität: Unternehmen in einem dynamischen Umfeld erfolgreich führen*. München: Haufe.

Lamerre, E. *et al.* (2023) *Rewired: The McKinsey Guide to Outcompeting in the Age of Digital and AI*. New Jersey: Wiley.

Lauer, T. (2019) *Change Management: Der Weg zum Ziel*. Wiesbaden: Springer.

Machado, C. and Davim, J.P. (2023) *Management for Digital Transformation*. Cham: Springer.

Mahidhar, V. and Davenport, T.D. (2018) 'Why Companies That Wait to Adopt AI May Never Catch Up', *Harvard Business Review*. Available at: https://hbr.org/2018/12/why-companies-that-wait-to-adopt-ai-may-never-catch-up (Accessed: 4 October 2023).

Park, Y.W. (2022) *Ambidextrous Global Strategy in the Era of Digital Transformation*. Cham: Springer.

Rogers, D.L. (2023) *The Digital Transformation Roadmap: Rebuild Your Organization for Continuous Change*. eBook. New York: Columbia University Press.

Schallmo, D.R.A. *et al.* (2023) *Digitalisierung: Fallstudien, Tools und Erkenntnisse für das digitale Zeitalter (Schwerpunkt Business Model Innovation)*. Wiesbaden: Springer.

Sendler, U. (2020) *KI-Kompass für Entscheider: Künstliche Intelligenz in der Industrie—Strategien, Potenziale, Use Cases*. München: Hanser Verlag.

Vagadia, B. (2020) *Digital Disruption: Implications and Opportunities for Economies, Society, Policy Makers and Business Leaders*. Cham: Springer.

Waller, S. (2016) *The Digital Champion: Connecting the Dots Between People, Work and Technology*. Online book.

Wittenstein, D. (2022) *Managing Digital Transformation: Evidence from Hidden Champions and Measurement Approaches (Innovation und Entrepreneurship)*. Wiesbaden: Springer.

Wolan, M. (2020) *Next Generation Digital Transformation: 50 Prinzipien für erfolgreichen Unternehmenswandel im Zeitalter der Künstlichen Intelligenz*. Wiesbaden: Springer.

Zaoui, F. and Souissi, N. (2020) 'Roadmap for Digital Transformation: A Literature Review', *Procedia Computer Science*, 175, pp. 621–628. The 7th International Conference on Emerging Internetworks, Communication and Mobility (EICM), 9–12 August, Leuven, Belgium. Available at: https://www.sciencedirect.com/science/article/pii/S187705092031447X.

Chapter 3

Smart Factories and Industry 4.0

3.1 Scope and Definition

"Smart factory" and "intelligent factory" are terms from operations management and lean management in the field of digitalised manufacturing technology. It is part of the German federal government's high-tech strategy as part of the Future Industry 4.0 project. It describes the vision of a production environment in which manufacturing plants and logistics systems largely organise themselves without human intervention in order to produce the desired products (Helmold, 2024).

A smart factory is a manufacturing solution that provides flexibility with lower costs and adaptive production processes, solving some of the problems that arise at production facilities in a world of increasing complexity. This manufacturing solution is based on automation – a combination of software, hardware, and mechanics – which leads to the optimisation of production processes, reducing unnecessary labour and wastage of resources. The smart factories are pioneering examples of a new era of lean and digital production methods. They are the culmination of the digital era of the past 20 years meeting the age-old practices of manufacturing. Smart factories illustrate the growing necessity for smarter production processes, smarter products, and hopefully an even smarter future.

Figure 3.1 shows the smart factory elements incorporating intelligent robots, automatic guided vehicles (AGVs), the Internet of Things

Figure 3.1. Smart factory elements.

(IoT), additive and flexible manufacturing, integrated supply chains, smart maintenance, Big Data, predictive analytics, realities, and cybersecurity in planning, production, and after-service across all value chain stages.

3.2 Digital Twin

A digital twin is a digital representation of a physical object, person, or process, contextualised within a digital version of its environment. Digital twins can help an organisation simulate real situations and their outcomes, ultimately allowing it to make better decisions (Stjepandic *et al.*, 2023). In a smart factory, machines, tools, resources, products, and storage systems receive a digital representation, the so-called “digital twin”. The condition of all components can be viewed in real time at any time via the digital twin representations. The so-called cyber-physical systems (CPSs) are created from mechanical and electronic components that are wirelessly networked with one another and communicate in an IoT or Industrial Internet of Things.

This communication is ensured by built-in chips in workpieces and production systems that are connected to the internet for data exchange. Workpieces to be processed become smart products by providing data about their own manufacturing process during manufacture, for example, by storing the data on an RFID chip. These data are used to control the path of the smart products through the system and individual production stages. Thanks to the mass of data, the production systems learn to make decisions independently as AI, reacting to unforeseen events and solving complex tasks. In this scenario, data become more of an economic asset and competitive advantage than ever before. The aim is to use the collected data and findings in such a way that, above all, the individual production of individual items tailored to customers is not only economically efficient and profitable but also suitable for mass customisation, making, for example, customer-specific mass customisation production possible (Rückert *et al.*, 2023).

3.3 Smart Factory Process

The smart factory process is described in Figure 3.2. The process integrates all participants in the value chain from suppliers to end customers. In smart factories, suppliers are connected with planning, production, and other processes through cloud systems and enterprise resource planning systems (ERPs), so that suppliers can anticipate and ideally balance production with actual customer needs. Logistics processes and partners are linked through real-time tracking and location information. It is necessary to connect machines, products, and people with real-time data. Big Data management supports these processes. Within the smart factory, robots and AGVs are used as part of the production process. Realities (augmented, virtual, and mixed), additive manufacturing methods and integrated cells are used to optimise processes and production steps. Smart quality management tools and statistical process control enrich the process up to customer delivery to secure customer satisfaction (Helmold, 2024).

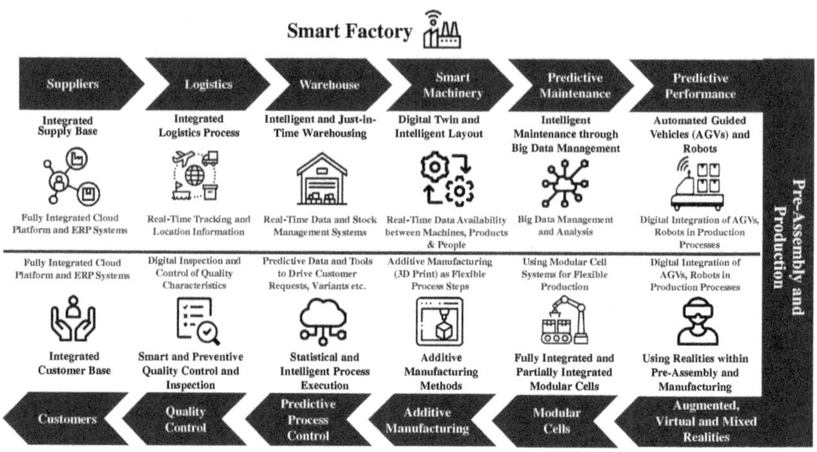

Figure 3.2. Smart factory process.

3.4 Areas Using Smart Factory Methods

3.4.1 Examples and areas

The networking of embedded and production systems and dynamic business and engineering processes enables profitable production of products, even for individual customer requirements up to batch size one. The technical basis is CPSs, which encompasses both physical production objects and their virtual image within a centralised system (see also digital twin) (Kletti and Rieger, 2023). In a broader context, the IoT is often mentioned. Part of this future scenario continues to be communication between the product (e.g., workpiece) and the production system: The product itself brings its production information in a machine-readable format, e.g., on an RFID chip. These data are used to control the product's path through the production system and the individual production stages. Other transmission technologies, such as WLAN, Bluetooth, colour coding, or QR codes, are currently being experimented with. At universities and research institutions, work is being done on the smart factory as part of so-called model factories, such as ARENA2036. Examples from industry include automobile production. For example, the "Factory 56" from the Mercedes-Benz Group is equipped with Wi-Fi and 5G networks. Various software applications network the machines and systems, enabling Big Data analyses, for example. The automobile manufacturer produces the Mercedes-Benz S-Class model in this factory and, from 2021, the electric variant

Mercedes-Benz EQS (Günnel, 2020). In connection with the smart factory, similar concepts are now also being developed in areas directly linked to product manufacture. For example, systems are being developed, especially in transportation and internal logistics and material flow, that can move and transport materials autonomously, thereby enabling autonomous production supply. Communication and location and identification technologies such as the internet and WLAN play a role here; GPS and RFID also play an important role.

3.4.2 The Tesla Gigafactory near Berlin

A better visualisation of smart factories can be found in the example of Tesla's Gigafactory at Grünheide, Berlin. This is the fourth Gigafactory under construction, following the ones built in Nevada, Buffalo, and Shanghai. It also happens to be their first Gigafactory in Europe and could prove to be Tesla's biggest factory yet. The Berlin Gigafactory, launched in 2021, will be the first of its kind, with Tesla promoting it as the most advanced high-volume electric vehicle production plant in the world. It will be used for building batteries, powertrains, and vehicles, starting with Model Y and Model 3 on a 300-hectare site in Grünheide, aiming to achieve a future production volume of 500,000 units annually. Tesla is striving to not only create a car that is smart but also to use smart means of production. The initial pictures of the factory released by company CEO Elon Musk depict their vision of a smart factory with solar panels dominating the rooftop, creating a more sustainable means of production. Its complex systems of producing cars have made headlines in the past, but the Berlin Gigafactory will take that up a notch. On its official website, Tesla mentions the use of a new dimension of casting systems and an efficient body shop, pushing forward the boundaries of vehicle safety. The paint shop introduces a new generation of colour tone depth and complexity. Tesla's constant drive to transform how factories function has allowed them to revolutionise the car industry.

3.4.3 Adidas speed factory in Ansbach

The desire to create the factory of the future led Adidas to build a speed factory in Ansbach, Germany. With robots assisting humans, the company focused on sneaker-production processes in a single space, thereby completing production in a few days and proving their speed theory. Focusing on mass customisation with shorter lead times, they use 3D printing technology to easily create digital replicas or mock-ups. Prototypes are quickly printed, facilitating an appropriate response to shifting consumer demands and satisfying customers' needs within days. Adidas may have shut down in Ansbach, yet they continue to use the speed factory technologies in the production of their athletic shoes in Asia. The company claims to have reinvented manufacturing in the footwear industry by replacing human hands in countries like Indonesia, Vietnam, and China by adopting technologies such as 3D printing, robotic arms, multi-layered particle machines, laser-cutting robots, and computerised knitting to make running shoes. They now aim to explore the use of 4D technology and other options to modernise production.

3.4.4 The connected factory of Bosch in Blaichach

Global manufacturing of automotive technologies such as anti-lock braking systems (ABSs) and electronic stability programs (ESPs) requires efficiency in production, and Bosch, at its lead plant in Blaichach, is doing exactly that. Implementing cutting-edge technology of connected manufacturing, the company uses 20 tablets every day to track processes, inspect machines, and check parts, data which are overlooked by 150 machine operators at the plant. Bosch's performance tracker system detects cycle-time deviations of mere milliseconds, enabling operators to react quickly and intervene at the earliest stage possible. Furthermore, an operator support system displays errors and provides recommendations on how to fix them (Bosch, 2019). Yet another technology is Bosch's Nexeed, a smart software suite that helps ensure production is running smoothly at their facilities. The company's connected industry cluster developed this software, which reads out data from 60,000 sensors and delivers relevant information to the appropriate employee in a timely manner. The team thereby keeps track of manufacturing operations in real time, making predictive maintenance possible. Bosch's worldwide manufacture of ABS/ESP safety systems has seen a productivity jump of 24 percent, thanks to such skilful innovations.

3.4.5 Infineon's smart factory in Dresden

The smart factory set-up of Infineon at Dresden wows the world with its intelligent networked manufacturing. The plant has achieved an automation level of 92 percent, with over 200 robots assisting employees. Infineon manufactures over 400 different products based on 200 and

300 mm wafers for all four segments of the group, quickly and to excellent quality. The 200 mm line is the world's most automated factory, while its 300 mm line was set up to achieve fully automated production, improving its productivity by 70 percent.

The company's system controls, wafer transport, and production management are linked to each other and controlled in real time via information technology (IT) systems. Furthermore, the systems can communicate with other Infineon sites around the world. Transport of the wafers is performed automatically, and the smart factory is controlled via central operating and monitoring systems as well as algorithms. The company optimises production control by simulating the impact of changes to its product portfolio in advance. This gives it the freedom to flexibly respond to its customers' needs. Infineon continues to develop new products and solutions for the automotive and power electronics industries while investing in the latest technologies, including AI, attracting attention for all the right reasons.

3.5 Challenge of Integration of Suppliers into the Smart Factory System

Specialised and global supply networks create complex supply networks. The challenge is here to integrate the supply base smoothly and smartly into the own value chain by intelligent and digital applications and ERPs. A method of short interval technology (SIT) can be used to systematically set up fast control loops in the company, including stakeholders and suppliers. The goal is perfect production with transparent, responsive, and economical processes (Kletti and Rieger, 2023). With the traditional methods of production planning and control, alongside frequently encountered production processes, it is only with great effort that many companies are able to operate effectively within increasingly networked and fast-paced supply chains, achieving short delivery times, high adherence to delivery dates, small batch sizes, short-term call-offs, and just-in-time concepts to ensure on-time deliveries at ideal and competitive costs.

3.6 Advantages of Smart Factories

It is not a surprise that the smart factory is also called the factory of the future. Digital change has and will continue to have a major impact on industry and society in general. The smart factory model enables companies to keep up with change and implement customers' new needs for individuality, flexibility, and speed. The advantages of using smart factories include the following:

- Lean and optimised processes with shorter production times and lower costs, which lead to increased productivity.
- Customer-oriented, individualised production of individual pieces up to batch size one at the cost of mass production.
- Automated and efficient ordering processes and supply chains by using predictive analytics or Big Data.
- Fast and agile adaptation to market requirements based on real-time information.
- Intelligent and transparent supply chain integration from suppliers to customers.
- Greater flexibility throughout the value chain with flexible manufacturing cells or additive manufacturing.
- Fast reactions to new or changed product and market requirements.
- Smart, digital, and preventive total productive maintenance process throughout the value chain.

3.7 Industry 4.0: Scope and Definition

Industry 4.0 refers to the intelligent networking of machines and processes in industry with the help of information and communication technology. Industry 4.0 offers great opportunities for Germany as a leading international industrial location. Around 15 million jobs depend on the manufacturing economy. They make a decisive contribution to the international competitive strength of German industry. In order to maintain and expand this competitive strength, it is important to tap into the enormous potential of Industry 4.0 at an early stage and to actively shape the digital structural change in the industry. Germany has a good starting position, particularly thanks to its competitive strength and expertise in the area of embedded systems. In this way, the potential of Industry 4.0 can be optimally used (BMWi, 2023). There are many ways for companies to use intelligent networking (Reinhart, 2017). The possibilities include, for example, flexible production systems, convertible factories, a customer-centric orientation, optimised logistics systems, or Big Data management (Helmold, 2023, 24).

- *Flexible production*: Many companies are involved in the production of a product and contribute step-by-step to its creation. Digitally networked, these steps can be better coordinated, and the utilisation of the machines can be better planned.

- *Convertible factory*: In the future, production lines will be built in modules. They can be quickly assembled for a task. Productivity and cost-effectiveness are improved, and individualised products can be manufactured in small quantities at affordable prices.
- *Customer-centric solutions*: Consumers and producers are moving closer together. Customers can help design products according to their wishes; for example, elements of sneakers can be designed by themselves and adapted to the individual foot shape. At the same time, smart products that have already been delivered and are in use can send data to the producer. With the usage data, the producer can improve their products and offer customers new services.
- *Optimised logistics*: Algorithms calculate ideal delivery routes, and machines independently report when they need new material – smart networking enables an optimal flow of goods.
- *Use of data*: Data on the production process and the condition of a product are brought together and evaluated. Data analysis provides information on how a product can be manufactured more efficiently. Even more importantly, it is the basis for completely new business models and services. For example, lift manufacturers can offer their customers "predictive maintenance". Lifts are equipped with sensors that continuously send data about their condition. Wear and tear can be detected and addressed before it causes a failure.
- *Resource-saving circular economy*: Products are viewed across their entire life cycle based on data. The form in which materials can be recycled is already determined in the design.

3.8 From the First Industrial Revolution to Digitalisation

3.8.1 Subject and definition

The Industrial Revolution refers to permanent changes in economic and social conditions, as well as working conditions. Since development began in the second half of the 19th century, we are now in the fourth stage of the Industrial Revolution: the digitalisation of industry. The previous phases were all crucial to our current "smart manufacturing", and the future will be too.

3.8.2 Birth of the steam engine (Industry 1.0)

The first Industrial Revolution, or Industry 1.0, refers to the industrial age that began around 1800 with the first mass-produced machines. Back then, production facilities were gradually being developed and the machines were mainly powered by water and steam power. The steam engine was the first machine that increasingly found its way into factories and significantly advanced industrialisation. The first phase of industrial development created many new jobs, particularly in North America and Europe, such as in textile mills, railroads, and coal mines.

3.8.3 Electricity as a crucial innovation in Industry 2.0

With the advent of electricity in the late 19th century, a second Industrial Revolution emerged (Industry 2.0). The assembly line, introduced to the automobile industry by Henry Ford in 1913, was particularly important: production was accelerated because each employee only completed one unit of work. Globalisation also progressed quickly, supported by aviation, as raw materials, food, and clothing were transported across wide national borders.

3.8.4 Invention of the computer (Industry 3.0)

The third Industrial Revolution (Industry 3.0) began in the 1970s and was characterised by IT and automation. While some large companies used computers in the 1940s, they are now becoming increasingly popular among individuals and the private sector. Compared to previous mass production in factories, human work is increasingly being taken over by machines.

3.8.5 Technological changes to Industry 4.0

The way we communicate is constantly changing, and the past few years have been marked by major digital shifts. While direct verbal communication remains important for most people, the exchange of information is increasingly taking place digitally. We are currently in an exciting phase characterised by technological upheaval and are ushering in a new era. As a result, the federal government came up with the idea of calling the fourth revolution in industrial development Industry 4.0 (Bogoviz *et al.*, 2018).

Figure 3.3 outlines the evolution from Industry 1.0 to Industry 4.0.

The First Industrial Revolution was marked by a transition from hand production methods to machines through the use of steam and water power. Its effects had consequences on textile manufacturing, which was the first to adopt such changes, as well as the iron industry, agriculture, and mining, although it also had societal effects with an ever-stronger middle class.

The Second Industrial Revolution, also known as the "Technological Revolution", was the period between 1871 and 1914 that resulted from the installation of extensive railroad and telegraph networks, allowing for faster transfer of people and ideas, as well as electricity. Increasing electrification allowed for factories to develop the modern production line. It was a period of great economic growth.

The Third Industrial Revolution, also known as the "Digital Electronics Revolution", occurred in the late 20th century following the end of the two World Wars, resulting from a slowdown in industrialisation and technological advancement compared to previous periods. The production of the Z1, which used binary floating-point numbers and Boolean logic, marked the beginning of further developments a decade later.

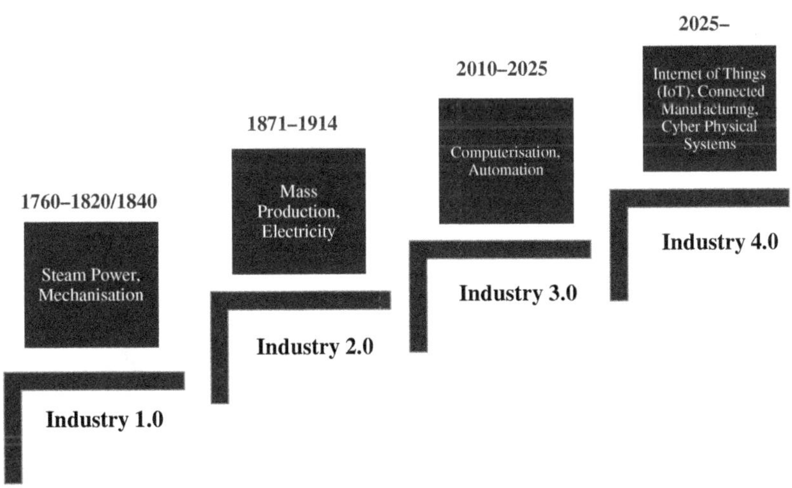

Figure 3.3. From Industry 1.0 to Industry 4.0.

In essence, the Fourth Industrial Revolution is the trend towards automation and data exchange in manufacturing technologies and processes, including CPSs, IoT, Industrial IoT, cloud computing, cognitive computing, and AI. The machines cannot replace the deep expertise, but they tend to be more efficient than humans in performing repetitive functions, and the combination of machine learning.

3.9 Increasing Productivity and Increasing Innovation through Industry 4.0

The introduction of Industry 4.0 is about a new way of organising and controlling the entire value chain of a product throughout its entire lifecycle. This cycle is increasingly oriented towards individualised customer requests and includes all stages from the idea to the concept, development, production, delivery to recycling, including the associated services (Moser, 2016). The goals of Industry 4.0 largely correspond to the goals of lean management. It is primarily about increasing competitiveness through:

- increasing productivity, even with a batch size of one;
- the self-configuration, self-diagnosis, and self-optimisation of systems;
- resource-saving and more efficient production methods;
- greater flexibility and responsiveness in response to increased customer requirements;
- the individualisation and customisation of products;
- driving innovation and the alignment of intelligent and sustainable business models with the digital economy.

3.10 Key Elements of Digitalisation

3.10.1 Machine-to-machine communication

Data and machine learning continue to drive digitalisation forward. This creates crucial pivotal points for the industry and fundamentally defines smart production processes (Heuser and Wimmer, 2023).

Machine-to-machine (M2M) refers to direct communication between devices over any communication channel, including wired and wireless communications. M2M communication can include industrial instrumentation that allows a sensor or measuring device to transmit the information it records (e.g., temperature, inventory levels) to application software that can use it (e.g., adjusting an industrial process based on temperature or placing orders to replenish inventory). Such communication was originally achieved by having a remote network of machines relay information back to a central hub for analysis, which was then forwarded to a system such as a personal computer. More recent M2M communication has evolved into a network system that transmits data to personal devices. The expansion of IP networks around the world has made M2M communication faster and easier while consuming less power. These networks also open up new business opportunities for consumers and suppliers.

3.10.2 Information transparency

In the future, it will not only be about what and how quickly companies produce and deliver but also how they produce the goods and services. This is because purchasing decisions are increasingly dependent on various criteria. This includes the use of green energies, the degree of CO_2 neutrality, and the extent of resource-saving processes and circular concepts. For example, through the use of recyclates and the development of smart recycling concepts, the conservation of resources should be progressively maximised. Environmentally friendly materials are used in the packaging, and all work stages and production processes are checked and constantly optimised to save valuable resources – such as water – and avoid hazardous substances. This is with the aim of establishing an effective circular economy that is ecologically and economically beneficial for all parties involved. When it comes to products and solutions, the automation specialist pays attention to sustainability in all process stages – from development to production and logistics to recycling. The focus is on durability, repair, and reuse. In the future, the ecological balance of products will be reflected in the product environmental footprint (PEF), which will most probably become a necessity from the coming years onwards. This creates a high level of transparency, for example, with regard to the CO_2 emissions of a product over its entire life cycle. The products and services manufactured in the factories constantly provide data about the current state. Suppliers, value creation partners, or workplaces that do not have interfaces can be integrated into the infrastructure via data collection boxes and AI.

3.10.3 Decentralised decisions

Many new and established companies are using AI to support the value proposition, value creation, value delivery, and revenue mechanisms. AI represents an innovation in the control of smart networks. Through decentralised use, AI should make networks even more effective in the future.

3.11 Industry 5.0

Industry 5.0, also known as the Fifth Industrial Revolution, is a new and emerging phase of industrialisation in which people work together with advanced technology and robots controlled by AI to improve workflows. This is accompanied by a greater focus on people, as well as increased resilience and an improved focus on sustainability (Helmold, 2024). This new phase, which encompasses more than just manufacturing, builds on the Fourth Industrial Revolution (Industry 4.0) and is made possible by developments in IT, including aspects such as AI, automation, Big Data analytics, the IoT, machine learning, robotics, intelligent systems, and virtualisation (Machado and Davim, 2023). While Industry 4.0 was still about optimising performance through networking production capacities (machines, processes, and systems) and the associated data exchange (Rot *et al.*, 2020), Industry 5.0 is all about changing paradigms and bringing resolutions since it will decrease emphasis on technology and assume that the potential for progress is based on collaboration between humans and machines. The ongoing developments are improving customer satisfaction by utilising and manufacturing personalised products. In modern business with paid technological developments, Industry 5.0 is required for gaining competitive advantages as well as economic growth for the factory (Adel, 2022). The key elements of Industry 5.0 are:

- human–robot collaboration,
- intelligent robots,
- smart machines,
- cognitive systems,
- dynamic resilience,
- personalisation of products,
- smart sustainability.

Industry 5.0 focuses on the collaboration between people and intelligent machines or robots (AI). Robots use technologies such as the IoT and Big

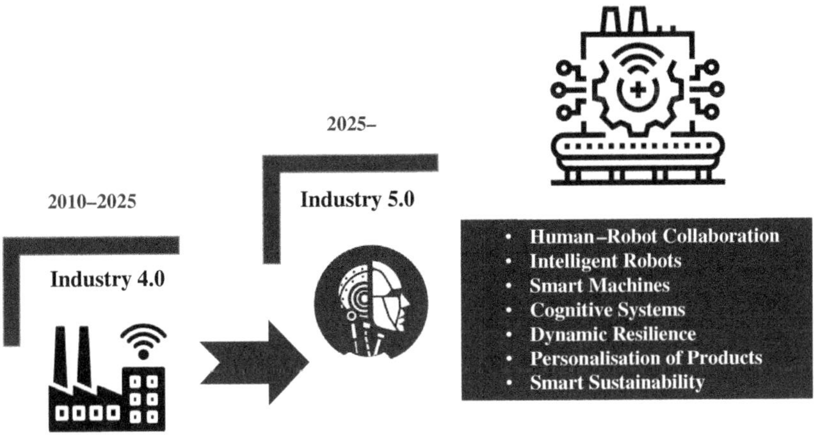

Figure 3.4. Industry 5.0.

Data to gain insights from available data and derive actions. The better humans and machines or robots interact with each other, the greater the gains in terms of efficiency and productivity (Adel, 2022). The associated increase in productivity has an epochal character. In addition, robots and machines are becoming more intelligent, so that products can be personalised quickly (Figure 3.4).

Bibliography

Adel, A. (2022) 'Future of Industry 5.0 in Society: Human-centric Solutions, Challenges and Prospective Research Areas', *Journal of Cloud Computing*, 11, Article number: 40.

BMWi (2023) 'Chancen durch Industrie 4.0: Von smarten Objekten und vernetzten Maschinen zurück zum Menschen'. Available at: https://www.plattform-i40.de/IP/Redaktion/DE/Standardartikel/chancen-durch-industrie-40.html (Accessed: 17 October 2023).

Bogoviz, A.V. *et al.* (2018) *Industry 4.0: Industrial Revolution of the 21st Century*. Wiesbaden: Springer.

Bosch (2019) 'Industrie 4.0: Smart Factory. Bosch stellt Weichen für die Fabrik der Zukunft'. Available at: https://www.bosch-presse.de/pressportal/de/de/bosch-stellt-weichen-fuer-die-fabrik-der-zukunft-195520.html (Accessed: 4 January 2024).

Budde, L. *et al.* (2023) *Smart Factory Navigator: Identifying and Implementing the Most Beneficial Use Cases for Your Company—44 Use Cases That Will*

Drive Your Operational Performance and Digital Service Business. Heidelberg: Springer.

Günnel, T. (2020) 'Mercedes-Benz eröffnet S-Klasse-Fabrik "Factory 56"', *Automobil-Industrie*, 5 January. Available at: https://www.automobil-industrie.vogel.de/mercedes-benz-eroeffnet-s-klasse-fabrik-factory-56-a-960628/ (Accessed: 5 January 2024).

Helmold, M. (2024) *Erfolgreiche Transformation zum digitalen Champion: Wettbewerbsvorteile durch Digitalisierung und Künstliche Intelligenz*. Wiesbaden: Springer.

Helmold, M. and Terry, B. (2021) *Operations and Supply Management 4.0: Industry Insights, Case Studies and Best Practices*. Cham: Springer.

Infineon (2023) 'Unleashing the Power of the IoT: One-stop Product Portfolio and Made-to-Measure Services for a Faster Time-to-Market. Smart Factory'. Available at: https://www.infineon.com/cms/de/about-infineon/make-iot-work/smart-factory/ (Accessed: 8 January 2024).

Kletti, J. and Rieger, T. (2023) *Die perfekte Produktion: Manufacturing Excellence in der Smart Factory*. Heidelberg: Springer.

Machado, C.F. and Davim, J.P. (2023) *Industry 5.0: Creative and innovative Organisation*. Cham: Springer.

Moser, D. (2016) 'Industrie 4.0: Einführung', *Wertfabrik*. Available at: https://www.wertfabrik.ch/de/blog/industrie-4-0/ (Accessed: 4 October 2022).

Rot, A. *et al.* (2018) *Towards Industry 4.0—Current Challenges in Information Systems*. Wiesbaden: Springer.

Rückert, T. *et al.* (2023) *Digital Twin Development: An Introduction to Simcenter Amesim*. Wiesbaden: Springer.

Stjepandic, J. *et al.* (2023) *DigiTwin: An Approach for Production Process Optimization in a Built Environment*. Wiesbaden: Springer.

Vogel-Heuser, B. and Wimmer, M. (2023) *Digital Transformation: Core Technologies and Emerging Topics from a Computer Science Perspective*. Wiesbaden: Springer.

Chapter 4

Automatic and Autonomous Driving Technology in the Mobility Sector

4.1 The Evolution of Autonomous Driving: A Century of Innovation

The concept of autonomous driving first captured public attention as early as in 1925 when Francis Houdina demonstrated a radio-controlled car in New York City. Dubbed the "phantom auto", the vehicle could start, shift gears, and brake without human intervention, sparking a mix of fascination and scepticism about the future of driverless technology (LaFrance, 2016).

In 1939, General Motors unveiled its visionary "Futurama" exhibit at the New York World's Fair, showcasing a bold concept of automated highways and self-driving cars integrated into an interconnected urban landscape (Tjokro, 2021). This vision was brought closer to reality in 1953, when General Motors and RCA collaborated to develop an automated highway system. Using embedded sensors and cables, this system provided a glimpse of how infrastructure could guide vehicles autonomously, laying the groundwork for the technologies that would later define autonomous vehicle (AV) systems (Ackerman, 2016).

The 1980s marked a shift from conceptualisation to implementation, with advancements in computing and robotics driving innovation. Carnegie Mellon University's NavLab and Mercedes-Benz's EUREKA Prometheus Project led the way in developing advanced autonomous systems. The NavLab project developed a series of AVs, with NavLab 5, a Pontiac Trans Sport, completing a cross-country trip from Pittsburgh to San Diego in 1995 (Carnegie Mellon University, n.a.). These pioneering initiatives, supported by DARPA, introduced key technologies such as obstacle detection, lane-keeping, and computer vision (Mobility21, 2024). In Europe, the German scientist Ernst Dickmann and his team at Bundeswehr University Munich developed AVs capable of navigating traffic at speeds up to 130 km/h. Their work included a successful demonstration in 1994, where two Mercedes 500 SELs autonomously navigated French commuter traffic. These early innovations laid the groundwork for modern AV technology, long before companies like Google, Tesla, and Uber entered the field (POLITICO, 2018).

The DARPA Grand Challenges of 2004 and 2005 marked pivotal moments in the evolution of AV development. In 2004, the inaugural event challenged teams to design fully AVs capable of navigating a 150-mile off-road course. While no team successfully completed the course, the competition catalysed significant technological advancements and highlighted the potential of autonomous systems. Building on this momentum, the 2005 challenge saw remarkable progress. Five teams successfully completed the course, with Stanford University's "Stanley" emerging as the winner after navigating a 132-mile desert route. This groundbreaking accomplishment demonstrated the feasibility of autonomous navigation, inspiring a surge of innovation and solidifying the DARPA Grand Challenge as a turning point in the journey towards modern AV technologies (DARPA, n.a.).

In 2009, Google launched its self-driving car initiative, which later evolved into Waymo, marking a significant milestone in the commercialisation of AV technology. This groundbreaking project demonstrated the capabilities of LiDAR-equipped vehicles to navigate complex environments and successfully complete autonomous routes, advancing both technological development and public awareness of autonomous driving (Waymo, 2020a).

By 2011, Nevada became the first U.S. state to legalise the testing of AVs on public roads, setting a regulatory precedent that accelerated global legislative advancements (Nevada DMV, 2023).

Automakers soon joined the race, with Audi demonstrating the RS 7 in 2015, a vehicle capable of high-speed autonomous manoeuvres, blending advanced driver-assistance systems (ADASs) with emerging autonomous capabilities (Motortrend, 2024).

In 2016, Uber took a significant step towards integrating AVs into everyday mobility by initiating tests of self-driving ride-sharing services in Pittsburgh. By 2020, Waymo reached a key milestone by launching a fully driverless taxi service in Phoenix, Arizona. This marked a breakthrough in public adoption and showcased the feasibility of AVs in commercial transportation (World Economic Forum, 2016; Waymo, 2020b).

The rapid evolution of autonomous driving is rooted in innovations across enabling technologies. Early systems such as basic cruise control have progressed to modern adaptive cruise control (ACC), electronic stability control (ESC), and collision avoidance systems (CASs), significantly improving vehicle safety and functionality (Xiao and Gao, 2010; Li *et al.*, 2013; Sanjana *et al.*, 2017). The key to these advancements is sensors such as radar, LiDAR, and cameras, which provide critical environmental perception capabilities. Complementing these are vehicle-to-everything (V2X) communication systems, enabling real-time data sharing for enhanced situational awareness and safe navigation in complex environments (Krasniqi and Hajrizi, 2016).

Introduced in 2014, Tesla's Autopilot system provides advanced driver-assistance features, including automatic steering, acceleration, and braking within a lane. Later on, the company offered the Full Self-Driving (FSD) package to include features such as Navigate on Autopilot, Auto Lane Change, and Traffic Light and Stop Sign Control. Tesla leverages vast amounts of data collected from its fleet of vehicles to continuously improve its autonomous driving algorithms based on advanced neural networks. This data-driven approach allows Tesla to refine its systems based on real-world driving conditions (Tesla, 2025).

At present, Chinese tech giants such as Baidu, Alibaba, and Tencent are heavily investing in AI and autonomous driving research, creating a competitive ecosystem that fuels innovation, with Baidu's Apollo Project providing an open platform for autonomous driving development and collaborating with various automakers (China-Britain Business Council, 2023). Leading Chinese automakers such as BYD and NIO are aggressively pursuing Level 3 (L3) ADAS capabilities, including features like adaptive cruise control, lane-keeping assistance, and automated parking. With supportive regulatory frameworks for autonomous driving from the Chinese government, companies such as AutoX and Pony.ai are driving forward commercial deployments of robotaxis in various cities, showcasing the practical applications of autonomous driving technologies (McKinsey, 2023).

In October 2024, Tesla unveiled its Cybercab robotaxi, designed to be fully autonomous, with no steering wheels or pedals. Alongside the Cybercab, the company also introduced the Robovan, a larger AV designed to transport more people or cargo (Yahoo, 2024).

These milestones underscore the rapid progress and ongoing efforts towards a future defined by autonomous and electric transportation. With continued innovation and collaboration among industry, governments, and researchers, advancements in AVs (AVs) and innovative business models such as shared autonomous fleets and subscription-based mobility services will reshape urban landscapes and set new standards for mobility worldwide.

4.2 SAE Levels of Driving Automation

SAE International, formerly known as the Society of Automotive Engineers, is a global professional association and standards development organisation established in 1905. Primarily focusing on the mobility and transportation sectors, the organisation serves key industries, including automotive, aerospace, and commercial vehicles in driving innovation and advancement of mobility technologies. SAE fulfils the above missions by providing globally recognised technical standards, professional certifications, and knowledge-sharing initiatives across diverse engineering disciplines (SAE International, 2025).

As the global automotive industry transitioned from conventional vehicles to ADASs and fully autonomous vehicles, inconsistent terminology and definitions posed significant challenges. The absence of a universal framework led to miscommunication among key stakeholders, including manufacturers, regulators, researchers, and the public. Terms like “autonomous”, “self-driving”, and “automated driving” were often used interchangeably, causing confusion about the actual capabilities and limitations of emerging technologies. To address this, standardised classification of automation levels became essential for improving technical communication, advancing academic research, and establishing robust traffic regulations and legal frameworks for autonomous driving (Wang *et al.*, 2021; Fraunhofer IKS, 2023).

Recognising the need, the SAE introduced the SAE J3016 standard in 2014. Officially titled "Recommended Practice: Taxonomy and Definitions for Terms Related to Driving Automation Systems for On-Road Motor Vehicles", this standard is commonly referred to as the SAE Levels of Driving Automation. In 2021, the SAE updated this standard in collaboration with the International Organization for Standardization (ISO), ensuring international consistency and global acceptance of the framework for the communication of AV technologies (SAE, 2021).

The SAE J3016 Standard provides a taxonomy for six levels of driving automation, from Level 0 (no automation) to Level 5 (full automation). Each level specifies the extent to which a vehicle can operate autonomously, outlining the roles of the driver and the system at each stage:

Level 0 (L0) – No Automation: At L0, the vehicle relies entirely on human control for all driving tasks, including steering, braking, accelerating, and environmental monitoring. Advanced driver assistance features, such as emergency braking, blind spot warnings, or lane departure alerts, may be present, but these systems are strictly passive. They assist drivers by providing warnings or minor interventions, such as applying braking force in emergencies, but they do not assume control of the vehicle. The driver remains fully responsible for all aspects of driving (SAE, 2021).

Level 1 (L1) – Driver Assistance: L1 introduces basic automation, where a single automated system supports the driver by controlling either steering or acceleration/deceleration but not both simultaneously. Examples include adaptive cruise control, which maintains a set speed and adjusts for nearby vehicles, or lane-centering assistance, which provides slight steering adjustments to stay within the lane. These systems function under specific conditions and require the driver to maintain full responsibility, remain attentive, and be prepared to take control at any time (SAE, 2021).

Level 2 (L2) – Partial Automation: Vehicles at L2 can manage both steering and acceleration/deceleration simultaneously, integrating these functions for smoother operation. Examples include highway driving assist systems or automated parking features that combine multiple driver assistance technologies. However, L2 systems are limited to specific scenarios, such as highways, and require continuous driver supervision. The driver must remain alert and ready to intervene immediately if the system encounters a situation it cannot handle (SAE, 2021) (Fraunhofer IKS, 2023).

Level 3 (L3) – Conditional Automation: At L3, vehicles are capable of monitoring their surroundings and performing all driving tasks under specific conditions, such as in stop-and-go traffic or geofenced urban environments. While the system is operating, drivers are permitted to engage in non-driving activities, such as checking their phones or reading. However, the system can request driver intervention when it encounters situations beyond its operational design domain (e.g., unexpected road conditions). Features such as traffic jam chauffeurs exemplify this level, requiring drivers to be available to retake control when prompted (SAE, 2021).
Level 4 (L4) – High Automation: Vehicles at L4 can perform all driving tasks autonomously without requiring human intervention in predefined conditions or environments. Examples include autonomous shuttles operating in urban centres or robotaxis navigating designated highway routes. These vehicles are designed to function independently in situations where the system's operational domain is clearly defined and controlled. Passengers at this level can fully disengage from driving, focusing on non-driving activities such as sleeping or working. Some L4 vehicles may retain optional human controls, such as steering wheels and pedals, though these are not essential for operation (SAE, 2021; Fraunhofer IKS, 2023).
Level 5 (L5) – Full Automation: L5 represents the ultimate vision of autonomous driving, where vehicles are capable of handling all driving tasks under all conditions, including extreme weather and complex urban scenarios such as navigating intersections. At this level, the vehicle fully replaces the human driver, eliminating the need for any manual controls, such as steering wheels or pedals. Passengers are entirely free to engage in any activity or simply relax during the journey. The achievement of L5 automation will require overcoming significant technological, ethical, and regulatory challenges, as it necessitates a level of intelligence and adaptability comparable to human decision-making (SAE, 2021; Wang *et al.*, 2021; Fraunhofer IKS, 2023).

The SAE Levels of Driving Automation provide more than a mere classification framework; they serve as a comprehensive reference for evaluating the feasibility, safety, and ethical considerations of AV technologies. By defining the roles of both human drivers and automated systems, the taxonomy facilitates the development of industry standards, regulatory policies, and public understanding. By offering a universally recognised standard, the SAE Levels of Driving Automation play a pivotal role in shaping the future of mobility, fostering innovation, and facilitating the seamless integration of AVs into society (Wang *et al.*, 2021; Fraunhofer IKS, 2023).

4.3 How Autonomous Driving Changes Our Lives

4.3.1 The revolution of connectivity

The rapid evolution of autonomous driving technology is reshaping multiple industries, from passenger vehicles and ride-hailing services to logistics, freight, and public transportation.

In passenger vehicles, AVs enhance safety by reducing human error and improve convenience by allowing passengers to reclaim time during commutes (Bosch, 2024; Garikapati and Shetiya, 2024). Features such as adaptive cruise control, lane-keeping assistance, and traffic jam assist transform driving experiences, while inter-vehicle communication optimises traffic flow and reduces congestion. These advancements also make transportation more accessible to underserved populations, including the elderly and physically challenged (McKinsey, 2023; Dathe, 2024).

In the ride-hailing sector, companies such as Waymo, Tesla, Cruise, and Baidu Apollo are pushing the boundaries of autonomous mobility. Waymo's achievements include launching the first commercial autonomous ride-hailing service (Waymo, 2024), while Tesla has integrated its FSD system into its vehicles, despite regulatory challenges. Cruise and Baidu Apollo have also made strides, with Cruise providing driverless taxi rides and Baidu Apollo launching its robotaxi service across Chinese cities (Cruise, 2024; Baidu, 2022). Each company brings a unique approach to autonomous technology, from safety-first designs to AI-driven solutions (Reuters, 2022).

In logistics and freight, autonomous trucks and drones are revolutionising supply chains by increasing efficiency, reducing costs, and improving delivery times (Xu and Duffy, 2021; Courcoubetis *et al.*, 2024; Dathe, 2024). These technologies enable 24/7 operations, optimised routing, and faster deliveries, addressing challenges such as labour shortages and rising fuel costs. Drones, in particular, are redefining last-mile delivery, offering precision and convenience, especially in urban areas (Xu and Duffy, 2021).

AI plays a critical role in the success of AVs by powering perception, decision-making, and control. Machine learning, sensor fusion, and real-time data processing allow AVs to navigate complex environments and ensure safety (Wang *et al.*, 2021; Fraunhofer IKS, 2023). Advanced communication technologies such as 5G and DSRC enable seamless data exchange between vehicles and infrastructure, while high-definition mapping materials guide AVs in dynamic environments. As these technologies continue to develop, they promise to transform mobility, logistics, and urban transportation on an unprecedented scale (Krasniqi and Hajrizi, 2016; Fraunhofer IKS, 2023).

4.3.2 Emerging business models

In October 2024, Elon Musk, the CEO of the automotive giant Tesla, announced the company's shift of focus to developing AVs, including the Robotaxi (Reuters, 2024). This event marked a turning point in the history of automotive history. One of the most immediate economic benefits of autonomous driving lies in cost reduction and efficiency improvements across urban transportation systems. By eliminating the need for human drivers, ride-hailing and robotaxi services can significantly reduce labour expenses, which traditionally constitute a major cost factor for taxi and ride-share operators.

Furthermore, AI-powered fleet management systems enable optimised vehicle deployment, ensuring that supply aligns with real-time demand while minimising unnecessary idle times. This results in fewer vehicles being required to serve a given number of passengers, ultimately leading to lower operational costs and reduced congestion on city streets. Beyond ride-sharing, autonomous trucks and delivery vehicles are expected to revolutionise logistics by increasing efficiency in freight transportation. Autonomous trucking can reduce shipping costs, improve delivery times, and minimise driver fatigue-related accidents. Real-time data analysis allows autonomous logistics fleets to optimise routes dynamically, avoid traffic bottlenecks, and reduce fuel consumption, making supply chains more cost-effective and environmentally sustainable (Yang, 2024).

Another key advantage of AVs is their ability to leverage advanced diagnostics and predictive maintenance technologies. Through continuous monitoring of vehicle performance, AI-driven maintenance systems can detect early signs of wear and tear, preventing costly breakdowns and reducing repair expenses. Predictive maintenance extends the operational lifespan of AVs by ensuring timely interventions, lowering downtime, and improving overall fleet reliability. This not only reduces the total cost of ownership for businesses operating AV fleets but also contributes to resource efficiency by minimising waste associated with vehicle replacements (Yang, 2024).

While the automation of driving-related tasks is expected to disrupt traditional employment in sectors such as taxi driving, trucking, and delivery services, it will also create new job opportunities in fields such as AI development, data analytics, vehicle maintenance, and smart infrastructure planning. As autonomous driving becomes more widespread, demand for specialised professionals in software engineering, cybersecurity, sensor calibration, and mobility management will rise. Additionally, cities will need to invest in the development and maintenance of smart infrastructure, such as intelligent roadways, real-time communication networks, and adaptive traffic management systems, further stimulating job creation in urban planning, construction, and digital technology industries (McKinsey, 2023).

The widespread adoption of AVs will necessitate significant upgrades to urban infrastructure. To maximise the benefits of AVs, cities must invest in intelligent transportation systems that enable real-time vehicle-to-infrastructure (V2I) and vehicle-to-vehicle (V2V) communication. These systems will enhance traffic flow, reduce congestion, and improve overall road safety by allowing AVs to anticipate and react to changing road conditions more effectively. Additionally, smart intersections equipped with AI-driven traffic signals can dynamically adjust to optimise vehicle movement, reducing unnecessary stops and improving fuel efficiency (Krasniqi and Hajrizi, 2016).

Parking infrastructure will also undergo a transformation. With autonomous ride-sharing services reducing private car ownership, demand for traditional parking facilities may decline, leading to repurposed urban spaces for public use, green areas, or commercial development. AVs will be capable of efficiently navigating and utilising designated drop-off and pick-up zones, reducing traffic congestion caused by circling vehicles looking for parking spaces (Krasniqi and Hajrizi, 2016).

4.3.3 Environmental implications

AVs contribute directly to sustainability by improving traffic efficiency. Autonomous systems use real-time data processing, V2V communication, and predictive algorithms to optimise driving patterns. This enables smoother traffic flow, reducing stop-and-go driving, which is a major contributor to fuel wastage and emissions in urban areas. Studies suggest that AVs can lower fuel consumption by minimising inefficient braking and acceleration patterns common among human drivers (Dathe *et al.*, 2024; Krasniqi and Hajrizi, 2016).

Additionally, AI-powered navigation allows AVs to select the most energy-efficient routes, avoiding high-traffic areas and minimising unnecessary travel. By reducing traffic congestion and enhancing road capacity, AVs can significantly decrease urban air pollution and energy waste (Dathe *et al.*, 2024; Krasniqi and Hajrizi, 2016).

Many AV prototypes and commercial deployments are designed to be fully electric, eliminating tailpipe emissions associated with internal combustion engine (ICE) vehicles. This transition aligns with global efforts to decarbonise transportation and reduce reliance on fossil fuels. However, the extent to which AVs contribute to emissions reductions depends on the energy mix used for charging. The environmental benefits are maximised if the electricity that powers AVs is generated from renewable sources such as wind, solar, or hydroelectric power (Krasniqi and Hajrizi, 2016).

Beyond electrification, AVs contribute to sustainability by improving fuel efficiency and reducing resource consumption. Compared to human drivers, AVs are programmed to operate within optimal efficiency parameters, without inefficient driving patterns such as sudden acceleration or hard braking. This precision minimises energy waste and extends the lifespan of critical vehicle components such as brakes, tyres, and engines. As a result, the reduction in wear and tear leads to fewer repairs and replacements, lowering overall resource consumption. These cumulative effects make AVs not only more energy-efficient but also less resource-intensive in terms of long-term maintenance (Dathe *et al.*, 2024; Courcoubetis *et al.*, 2024; Silva *et al.*, 2022).

Despite the potential benefits of autonomous driving, its environmental impact must be assessed holistically, taking into account the resource-intensive production of specialised AV components. The manufacturing of LiDAR systems, high-performance sensors, and onboard computing units involves energy-intensive processes that generate emissions and require rare earth materials. These environmental costs may offset some of the operational gains achieved through more efficient driving. In addition, battery technology also plays a crucial role in the environmental footprint of AVs. The production of lithium-ion batteries involves resource-intensive mining processes, particularly for materials such as lithium, cobalt, and nickel. Sustainable battery development – including advances in solid-state batteries, battery recycling, and second-life applications – will be critical in reducing the lifecycle emissions of electric AVs (Krasniqi and Hajrizi, 2016).

4.4 Future Trends and Further Considerations

Autonomous driving is set to redefine transportation by enhancing safety, optimising traffic flows, and minimising carbon emissions. As this transformative technology advances, it has become a strategic priority for governments, investors, and industry leaders worldwide. The urgency to capture this rapidly evolving market is driven not only by the substantial economic benefits but also by the broader societal advantages, such as improved mobility for underserved populations, enhanced accessibility for elderly and physically challenged individuals, and the development of smarter, more efficient urban environments. Companies that successfully integrate technological leadership with forward-thinking business strategies will be best positioned to thrive in the autonomous driving era (McKinsey, 2023).

To maximise the sustainability potential of AVs, a systemic approach is required. Primarily, the deployment of AVs on a large scale necessitates investments in digital infrastructure, such as 5G communication networks, high-definition mapping systems, and advanced traffic management platforms, all of which have their own environmental footprints (Krasniqi and Hajrizi, 2016).

The rise of AVs is also reshaping business models, shifting from private car ownership to shared autonomous fleets, such as robotaxis, autonomous shuttles, and mobility-as-a-service (MaaS). At present, pioneers such as Baidu and Tesla are leading efforts in robotaxi deployment, demonstrating the shift towards autonomous fleets.

Innovations in AI, V2X communication, and autonomous flying vehicles promise further disruptions in mobility, logistics, and urban planning. However, concerns around employment displacement, insurance liability, data privacy, and algorithmic transparency must be addressed (Dathe *et al.*, 2024; Silva *et al.*, 2022). Additionally, geopolitical competition between the U.S. and China highlights the strategic importance of AV technology.

Policymakers and industry players must intensify their collaborations to balance innovation with sustainability, ensuring that autonomous mobility delivers maximum societal benefits while mitigating risks.

Bibliography

Ackerman, E. (2016) 'Self-driving Cars were Just Around the Corner—in 1960', *IEEE Spectrum*, 31 August. Available at: https://spectrum.ieee.org/selfdriving-cars-were-just-around-the-cornerin-1960 (Accessed: 3 February 2025).

Baidu (2022) 'Baidu to Add 200 Fully Driverless Robotaxis to Fleet in 2023, Aiming for World's Largest Fully Driverless Ride-hailing Area', *PR Newswire*, 29 December. Available at: https://www.prnewswire.com/news-releases/baidu-to-add-200-fully-driverless-robotaxis-to-fleet-in-2023-aiming-for-worlds-largest-fully-driverless-ride-hailing-area-301711296.html (Accessed: 3 February 2025).

Bosch (2024) 'Automated Driving'. Available at: https://www.bosch.com/stories/topics/automated-driving/ (Accessed: 3 February 2025).

Carnegie Mellon University (n.d.) 'From 0-70 in 30'. Available at: https://www.cmu.edu/homepage/environment/2014/fall/from-0-70-in-30.shtml/ (Accessed: 3 February 2025).

China-Britain Business Council (2023) 'The Future of Autonomous Vehicles in China', *China-Britain Business Focus*, 11 September. Available at: https://focus.cbbc.org/the-future-of-autonomous-vehicles-in-china/ (Accessed: 3 February 2025).

Courcoubetis, C. *et al.* (2024) 'The Impact of Ride-Hailing in City Transportation ', *Frontiers in Engineering Management*. doi: 10.1007/s42524-024-4198-0.

Cruise (2024) 'History of Cruise'. Available at: https://getcruise.com/about/ (Accessed: 3 February 2025).

DARPA (n.d.) 'What is the Urban Challenge?', *Urban Challenge*. Available at: https://www.grandchallenge.org/grandchallenge/overview.html (Accessed: 3 February 2025).

Dathe, T. (2024) 'Automated Driving and AI in Mobility' in Helmold, M., Yilmaz, A.K. and Dathe, T. (eds.) *The Digital Champion: Best-practices and Insights for the Successful Transformation and Digitization Roadmap*. Singapore: World Scientific Publishing.

Dathe, T. *et al.* (2024) *Implementing Environmental, Social and Governance (ESG) Principles for Sustainable Businesses: A Practical Guide in Sustainability Management*. Cham: Springer.

Fraunhofer IKS (2023) 'Autonomous Driving'. Available at: https://www.iks.fraunhofer.de/en/topics/autonomous-driving.html (Accessed: 3 February 2025).

Garikapati, D. and Shetiya, S.S. (2024) 'Autonomous Vehicles: Evolution of Artificial Intelligence and the Current Industry Landscape', *Big Data and Cognitive Computing*, 8(4), 42. doi: 10.3390/bdcc8040042.

Krasniqi, X. and Hajrizi, E. (2016) 'Use of IoT Technology to Drive the Automotive Industry from Connected to Full Autonomous Vehicles', *IFAC-PapersOnLine*, 49(29), pp. 269–274. doi: 10.1016/j.ifacol.2016.11.078.

LaFrance, A. (2016) 'Your Grandmother's Driverless Car', *The Atlantic*, 29 June. Available at: https://www.theatlantic.com/technology/archive/2016/06/beep-beep/489029/ (Accessed: 3 February 2025).

Li, L. *et al*. (2013) 'Progress on Vehicle Dynamics Stability Control System', *Journal of Mechanical Engineering*, 49(24).

McKinsey (2023) 'Autonomous Driving's Future: Convenient and Connected', 6 January. Available at: https://www.mckinsey.com/industries/automotive-and-assembly/our-insights/autonomous-drivings-future-convenient-and-connected (Accessed: 3 February 2025).

Mobility21 (2024) 'The Road Ahead for Self-Driving Cars'. Available at: https://mobility21.cmu.edu/the-road-ahead-for-self-driving-cars/ (Accessed: 3 February 2025).

Motortrend (2024) '2015 Audi RS 7 Expert Review'. Available at: https://www.motortrend.com/cars/audi/rs-7/2015/ (Accessed: 3 February 2025).

Nevada DMV (2023) 'Autonomous & Alternative Electronic Transportation System Vehicles'. Available at: https://copilot.microsoft.com/chats/GxBzg2dYboYu8hnDBec1Z (Accessed: 3 February 2025).

POLITICO (2018) 'The Man Who Invented the Self-Driving Car (in 1986)', 19 July. Available at: https://www.politico.eu/article/delf-driving-car-born-1986-ernst-dickmanns-mercedes/ (Accessed: 3 February 2025).

Reuters (2022) 'GM Seeks U.S Approval to Deploy Self-Driving Vehicles', 18 February. Available at: https://www.reuters.com/business/autos-transportation/gm-seeks-us-approval-deploy-self-driving-vehicle-2022-02-18/ (Accessed: 3 February 2025).

Reuters (2024) 'Musk Now Says It's "Pointless" to Build a $25,000 Tesla for Human Drivers', *US News*, 10 October. Available at: https://www.usnews.com/news/top-news/articles/2024-11-05/musk-now-says-its-pointless-to-build-a-25-000-tesla-for-human-drivers (Accessed: 3 February 2025).

SAE (2021) 'SAE Levels of Driving Automation™ Refined for Clarity and International Audience', 3 May. Available at: https://www.sae.org/blog/sae-j3016-update (Accessed: 3 February 2025).

SAE International (2025) 'About SAE International'. Available at: https://www.sae.org/about/history (Accessed: 3 February 2025).

Sanjana, T. *et al*. (2017) 'Automated Anti-Collision System for Automobiles', *International Conference on Electrical, Computer and Communication Engineering (ECCE)*. Cox's Bazar, Bangladesh: IEEE, pp. 16–18.

Silva, Ó. *et al.* (2022) 'Environmental Impacts of Autonomous Vehicles: A Review of the Scientific Literature', *Science of the Total Environment*, 830, 154615. doi: 10.1016/j.scitotenv.2022.154615.

Tesla (2025) 'Future of Driving'. Available at: https://www.tesla.com/autopilot (Accessed: 3 February 2025).

Tjokro, M. (2021) 'Why Self-Driving Cars: From Horseless to Driverless', *Mobility21 News*, 20 April. Available at: https://mobility21.cmu.edu/why-self-driving-cars-from-horseless-to-driverless/ (Accessed: 3 February 2025).

Wang, J. *et al.* (2021) 'Towards the Unified Principles for Level 5 Autonomous Vehicles', *Engineering*, 7(9), pp. 1313–1325. doi: 10.1016/j.eng.2020.10.018.

Waymo (2020a) 'In the Driver's Seat: Footage from our 2009–2010 1,000 Autonomous Mile Challenge', 22 April. Available at: https://waymo.com/blog/2020/04/in-the-drivers-seat-1000-mile-challenge/ (Accessed: 3 February 2025).

Waymo (2020b) 'Waymo is opening its fully driverless service to the general public in Phoenix', 8 October. Available at: https://waymo.com/blog/2020/10/waymo-is-opening-its-fully-driverless-service-in-phoenix/ (Accessed: 3 February 2025).

Waymo (2024) 'Starting at Mile One — Waymo Story'. Available at: https://waymo.com/about/#story (Accessed: 3 February 2025).

World Economic Forum (2016) 'Uber is Rolling Out Its First Self-Driving Cars (With Humans Behind the Wheel)', 19 August. Available at: https://www.weforum.org/stories/2016/08/uber-s-rolling-out-its-first-self-driving-cars-although-they-ll-still-have-humans-at-the-wheel/ (Accessed: 3 February 2025).

Xiao, L. and Gao, F. (2010) 'A Comprehensive Review of the Development of Adaptive Cruise Control Systems', *Vehicle System Dynamics*, 48(10), pp. 1167–1192. doi: 10.1080/00423110903365910.

Xu, Z. and Duffy, V. (2021) 'A Systematic Review of Autonomous Driving in Transportation' in *HCI International 2021 — Late Breaking Papers: HCI Applications in Health, Transport, and Industry*. Cham: Springer. doi: 10.1007/978-3-030-90966-6_28.

Yahoo (2024) 'Tesla's Cybercab Robotaxi is Finally Here With a $30K Price Tag—Plus a Surprise Robovan', *Yahoo! Finance*, 11 October. Available at: https://finance.yahoo.com/news/teslas-cybercab-robotaxi-is-finally-here-with-a-30k-price-tag--plus-a-surprise-robovan-071844079.html?guccounter=1 (Accessed: 3 February 2025).

Yang, Z. (2024) 'What's Next for Robotaxis in 2024', *MIT Technology Review*, 23 January. Available at: https://www.technologyreview.com/2024/01/23/1086936/whats-next-for-robotaxis-2024/ (Accessed: 3 February 2025).

Chapter 5

Cutting-Edge Natural Language Processing Technologies

5.1 Natural Language Processing: Modern Leader Assistant

The utilisation of natural language processing (NLP) and AI as assistants to leaders is a subject that merits further investigation (see Figure 5.1). The content and outputs, the decision-making mechanism, and the responses of systems such as AI and NLP are influenced by the team that creates the software. The leadership understanding and leadership structure of societies and the culture, values, and location of organisations (such as DeepSeek based in China, Gemini's creator Google and ChatGPT's creator OpenAI, a nonprofit, both based in the U.S., etc.) are also relevant factors. Given the distinct nature of the Western and Eastern schools of thought, the responses generated by the software are likely to vary in accordance with the shifting social perspectives. To illustrate this point, consider the scenario of a malfunctioning fighter jet. In such an event, a crash is an unavoidable consequence. In this scenario, the optimal location for a crash is a matter of strategic importance. The nature of the responses generated by the software is contingent upon the value system and cultural context of the pilot operating the fighter jet. The objective is to either crash the plane outside the settlement, save the plane at all costs, or leave the plane and save oneself (a cost–benefit analysis according to society). Consequently, intelligent systems generate divergent responses with regard to both risk prioritisation and value assignment. Consequently, the selection of the intelligent system's software by the leader should be informed by their personal values and preferences, and it should be made from among the available alternatives. It is evident that AI and its constituent subsystems are capable of generating responses that are influenced by the cultural, value-based, political, ideological, and strategic factors that are representative of the company. It is therefore incumbent upon the leader to select the most appropriate assistant to fulfil these responsibilities. Consequently, AI and NLP have the potential to function as assistants to leaders. This option would undoubtedly be ideal for risk leaders who possess the capacity to learn, aspire towards perpetual self-improvement, and can judiciously evaluate alternatives.

Since communication is the determining factor wherever there are people, including industries, every innovation related to and/or affecting communication spreads and spreads faster than others. In keeping with human nature, humans need communication the most. Through communication, the leader influences people, inspires them, and guides them. People also want to convey to their leaders that they are following him through all kinds of communication channels. People follow, comment, react, and share to get reactions. Therefore, communication is a

Figure 5.1. AI as assistant for leaders.

Source: Created using Napkin AI (https://www.napkin.ai).

comprehensive digital technology that can be appropriated by people before others: selectivity in perception.

In the past, since the most critical point in wars was communication, codes that could not be understood by enemies were developed and used in communication. But now, thanks to technologies such as NLP and smart systems, codes can be deciphered in a shorter time. In the past, a few digits were sufficient, but now it is desired to have at least 16 digits, including numerical and verbal characters, and formed by adding signs. Otherwise, the passwords will be easy to crack.

At this point, new digital technologies that companies can use to reach their customers faster, better, and clearer, as well as to analyse customer inputs, create a new transformation.

Businesses use NLP software and tools to simplify, automate, and streamline their operations efficiently and accurately. Businesses in the insurance, legal, and healthcare industries use NLP technology to process, sort, and organise large volumes of medical records, financial data, and private documents. For example, Chisel AI helps insurance companies extract data, such as policy numbers, expiry dates, and customer information, from unstructured documents using Amazon Comprehend. In customer relations, it is used to scale customer service by making chatbots and voice bots more human-like. PubNub uses Amazon Comprehend to deliver a localised chat experience to its global customers, while T-Mobile identifies keywords in customers' text messages to provide personalised recommendations. Oklahoma State University integrates machine learning (ML) and NLP technologies in a chatbot solution that answers students' questions. In marketing, businesses use NLP tools such as Amazon Comprehend and Amazon Lex to analyse customers' emotions towards products or services. The Success KPI helps businesses focus on targeted areas of sentiment analysis and gain actionable insights from call centre data (AWS Amazon, 2025).

Digital transformation encompasses the use of a variety of advanced technologies, including NLP and AI. These offer strategic opportunities for management and organisation, particularly within the context of today's business environment. However, as with any risk, if it is not managed effectively, it has the potential to become a threat to companies. It is important to note that an event is not considered as either a threat or an opportunity in isolation. In line with the yin–yang philosophy, it is necessary to analyse the events that carry opportunities and threats at different levels. The risk management function suggests that it is advisable to embrace events that present a greater opportunity than risk. In instances where threats are deemed more probable or have the potential to inflict significant harm, measures are implemented to either eliminate or mitigate the risk, if feasible. However, there is always a residual risk. Furthermore, each risk has a cost attached, meaning that effective management requires the allocation of resources. Access to resources for industries and companies is becoming more restricted. Therefore, it is imperative to make informed investment decisions regarding digital technologies by evaluating the potential risks and ensuring the optimal utilisation of resources.

While NLP presents opportunities that can create major transformations in the automotive, marine, and aerospace industries, it can also bring serious risks. Digital transformation momentum is high. Since it has not yet fully mastered its scope, industries, and companies, this makes optimum management unlikely. It will be useful to explain this with an example.

As is known, there have been two aircraft crashes in the U.S. recently. One of them is "A passenger plane collided with a Sikorsky military helicopter in mid-air while approaching Ronald Reagan Airport in Washington, the capital of the United States. During the search for the accident in which 67 people died, the black boxes of the plane and helicopter were found" (BBC News, 2925).

In recent years, NLP has made significant progress, especially with the introduction of transformer models and attention mechanisms. These innovations have improved how machines understand language by allowing them to better capture context. Models such as BERT and GPT-3, have shown strong performance in interpreting word meanings accurately within text. Contextual embeddings have also advanced tasks such as translation and summarisation by providing more flexible and accurate understanding compared to older methods that rely on fixed word meanings. Another major breakthrough is transfer learning, where models pretrained on large datasets are adapted to specific tasks. This approach has made NLP systems more efficient and expanded their range of applications (Pragada, 2024). These advancements have led to practical benefits across industries. In healthcare, NLP is used to extract critical information from clinical notes and support diagnostic decisions. In finance, it powers sentiment analysis of market news and automates compliance monitoring. In aviation, NLP helps analyse safety reports to predict potential hazards. In the maritime industry, NLP is increasingly used to process ship logbooks, analyse maintenance records, and support incident reporting, contributing to safer and more efficient operations. Within supply chain management, NLP tools enable real-time monitoring of logistics documents, detect bottlenecks through unstructured feedback, and help forecast disruptions. In lean management, NLP can analyse employee suggestions, quality incident narratives, and internal reports to identify recurring inefficiencies and support continuous improvement initiatives.

The available information about the recent plane crashes around the world remains insufficient to determine what caused them. At this point, NLP technologies can be used in the fields of communication, data processing, and data analysis: for example, to improve communication between the flight crew and air traffic control (ATC), process voice commands, and analyse flight data, etc.

Although digital transformation and high-tech systems influencing accidents and crashes, along with AI, might be suggested as the primary causes, there is no definite information about whether or not accidents have a direct connection with NLP systems. In fact, it is well-known that aviation accidents do not have a single cause but occur due to a chain of factors (mainly human factors). Among these causes, it is necessary to analyse whether issues originating from digital systems alone are the direct source of accidents or have a significant influence on the occurrence of accidents/crashes, as well as to examine official reports. A well-known fact is that digital systems can potentially be a source of risk. For example, a misinterpretation of a voice command can lead to unintended consequences. Therefore, just like all systems, the correct usage, reliability, and accuracy of NLP systems are critical factors to ensure their effectiveness and prevent errors.

Also, lean management focuses on improving efficiency and eliminating waste, and technologies such as NLP and digital twins can enhance these processes. NLP helps analyse text data quickly, such as employee feedback, customer reviews, and reports, identifying areas for improvement. Digital twins create virtual models of production and processes, allowing real-time data and simulations to identify bottlenecks and inefficiencies, helping to optimise operations. Together, these technologies support lean management's goals by enabling faster decision-making and continuous improvement.

What is known for sure is that digital technology and related systems are potentially a source of risk (both opportunities and threats). For example, a misunderstanding of a voice command can lead to unintended consequences. Therefore, like all systems, the correct use of NLP systems, system reliability and sensitive points are sensitive points.

One of the important points we pay attention to in risk management is reporting. Reporting is also useful at every stage of risk management in terms of producing outputs for the next stage. In addition, reporting is required at the end of all operations, such as operational activities (for example, maintenance and repair in aviation, ATC, and flight), as well as the audits carried out by relevant authorities, and the reports prepared to respond to these inspections are also parts of the process. At this point, the use of digital technologies in reporting is also beneficial in terms of reducing errors and deficiencies in reports. However, it should not be forgotten that if there are errors, subjectivities or deficiencies in the data or data sources used to create reports, this will be reflected in the linguistic patterns created by NLP in some way.

The utilisation of AI and NLP-powered predictive operation platforms (comprising flight, maintenance, repair and operations (MRO), ATC, simulators, and planning) enables leaders to make more informed decisions regarding the planning and implementation of operations and managerial functions. Airbus has utilised AI technologies in various domains for an extended period and has recently undergone a substantial expansion in the scope of its applications in this field, leveraging the opportunities afforded by generative artificial intelligence (GenAI). The expansion was initiated in 2016, with the processing of high-resolution satellite images on the OneAtlas platform, and subsequently evolved into maintenance systems that predict aircraft failures using NLP technologies through the Skywise digital platform in 2017. These technologies, which were initially accessible only to specialised engineers and data scientists, are now available to a much wider range of employees thanks to GenAI. The company-wide GenAI working group, established in 2023, aims to make large language models (LLMs) and base models available to Airbus in an ethical, secure, and sustainable manner. The company has identified more than 600 use cases, ranging from engineering assistants to cyber threat analysis, procurement optimisation, and recruitment tools. These technologies have the potential to improve internal processes and transform customer relationships. To illustrate this point, consider a chatbot capable of searching through technical documents containing production instructions using natural language. Such a capability would result in time savings for field teams and an increase in productivity. Airbus seeks to disseminate these innovations offered by GenAI in a responsible manner, in accordance with the regulations that govern the aviation industry (Airbus, 2024).

NLP has become a critical tool in aviation, with applications ranging from the analysis of safety reports to the prediction of future risks. NLP-based systems are used across different countries to support a wide range of functions, including digital tracking of maintenance activities and automated classification of incident reports. These tools not only strengthen decision support systems but also enhance operational safety and security (Paul *et al.*, 2018). The Aviation Safety Reporting System (ASRS) is one of the most well-known and long-standing voluntary incident reporting programmes in aviation. Founded and managed by NASA in 1976, this system has been collecting reports on aviation incidents that have been voluntarily submitted throughout the U.S. since its inception. European Coordination Centre for Accident and Incident Reporting Systems (ECCAIRS) is an initiative that aims to standardise the collection, sharing, and analysis of accident and incident data throughout the European Union. Developed by the European Commission's Joint Research Centre, the mission of this initiative is to help national and European transport authorities improve public safety by facilitating the exchange of safety information (Tanguy *et al.*, 2016).

The application of NLP has the potential to enhance the quality of reporting and reports by promoting data-driven decision-making, cultivating a robust safety culture, fostering international collaboration, and catalysing innovation in safety management systems. The integration of AI and NLP into safety management and reporting systems has the potential to enhance the efficiency and effectiveness of aviation safety management. The utilisation of AI and NLP has enabled the expeditious review of reports received during the day, in comparison to conventional manual methods. These systems have the capacity to play critical roles in shaping safety policies, driving data-driven decision-making processes, and fostering international cooperation. By encouraging voluntary reporting of aviation incidents, both systems can support leaders in the establishment of a robust safety culture. The application of AI and NLP technologies to the analysis of reports can facilitate decision-making processes by enabling the processing of substantial volumes of textual and structured data, which can then be analysed and interpreted with greater efficiency.

5.2 NLP Concept and Techniques

NLP technologies can receive feedback from internal and external customers, analyse the feedback received, and analyse the suggestions from Kaizen (in line with the philosophy of continuous improvement). The results of this analysis can also be grouped. These outputs in the form of analyses and groupings act as a decision support system, helping to optimise decision processes.

Despite the abundance of data at their disposal, human resources (HR) departments have historically exhibited a marked reluctance to transform these data into meaningful insights. However, the integration of AI, ML, and especially NLP technologies has begun to effect a change in this situation. NLP is a branch of AI that facilitates the comprehension of human language through the utilisation of algorithms. This technology has the potential to transform the field of HR by analysing both written and spoken communication. NLP facilitates the transition of HR into a data-driven and strategic function by addressing the intricacies of human language. The primary domains in which NLP is employed within the field of HR may be categorised into three overarching categories (Spiceworks, 2025):

- **NLP in Recruitment:** NLP analyses resumes quickly and without bias, ranks candidates, and speeds up the recruitment process with chatbots.
- **NLP in Employee Engagement:** Employee satisfaction is analysed from surveys and feedback, problems are identified, and interaction strategies are developed.
- **NLP in Social Media Analytics:** By analysing the social media content of employees, their potential, level of commitment, and behavioural tendencies are measured, which supports talent management.

NLP and AI encompass a wider range of human-like intelligence capabilities, while NLP focuses on language-related problems. NLP is a technology that enables computers to understand, process, and respond to human language. NLP is used to process text or speech to make sense of it, analyse the structure of the language, and produce a response accordingly (Hirschberg and Manning, 2015).

Comprehensive, multi-page, multi-sectional large-volume reports, including different fields, require a significant amount of long-term resource use to review the articles. The attention and focus of the people involved in this examination do not remain constant during the process.

Attention may be distracted, attention may be collected, focus may be made better, or focus may be lost. In cases where people are involved, this variability is part of the work and the result. But at the same time, all human characteristics are the basis of originality and self-talent.

NLP is a field of study that deals with the interaction between linguistics and AI, offering a transformation that enables machines not only to process language but also to understand and respond to it. NLP advances human–machine interaction by using techniques such as language modelling, syntactic parsing, and semantic understanding (Pragada, 2024). NLP techniques can contribute to decision-making processes by using various analytical methods such as computer-aided text analysis, supervised machine learning, and unsupervised machine learning. These tools enable leaders to analyse large volumes of text data related to their behaviour, communication styles, and decision-making processes.

At the very beginning of the process, while preparing reports containing these long and comprehensive texts, the nature of the report varies depending on their areas of expertise, interests, and reporting competencies. In order to keep this situation under control and guarantee a certain scope in industries, such as aviation, details such as the format, content and fields to be filled in the reports are specific, and these are not left to the user's preference. Despite this, the quality of a report is only as strong as the people who prepared it and the data used to prepare it.

NLP can be used to assess leaders' written or verbal communications, providing insights into their effectiveness and areas for improvement (Tonidandel and Albritton, Transforming Leadership Assessment Using Natural Language Processing, 2023). Built on NLP foundations, chatbots and virtual assistants can offer real-time feedback and support to leaders. These AI-powered tools can provide leaders with instant analytics and recommendations based on their interactions and behaviours, helping them improve their communication skills, decision-making processes, and overall effectiveness (Rajendran *et al.*, 2024).

5.3 The Interaction between NLP Technology and Leadership and Organisational Behaviour

Leadership is a multifaceted concept due to its diversity of styles, the sum of characteristics, and its effects on individuals, societies, and organisations. At its core, leadership involves influencing and guiding individuals or groups towards achieving common goals. Effective leadership is characterised by the ability to inspire and motivate followers, create a vision, and create an environment conducive to collaboration and innovation (Alsarrani and Jusoh, 2019; Ghasabeh *et al.*, 2015).

As a leadership style that supports adaptation with transformational leadership, flexibility and a contingency approach, it also positively affects the adoption of digital technology. The adoption of digital technologies also affects organisational agility and performance, as it affects and shapes organisational functions and the performance of operational activities. At this point, leadership that embraces and advocates for technological innovation can reshape organisations into volatile, dynamic work environments, thereby improving organisational capacity. From this point of view, it is revealed that digital technology plays a mediating role in the relationship between transformational leadership and corporate agility and performance (Yamin and Murwaningsari, 2023).

The relationship between NLP technology and leadership is one of development and transformation. NLP provides more objective and data-driven approaches to leadership assessment by introducing new methodologies for assessing and developing leadership skills. As NLP technology continues to evolve, it has the potential to revolutionarily change leadership training and development, offering personalised, real-time feedback and support to leaders in a variety of fields (Rajendran *et al.*, 2024; Tonidandel and Albritton, 2023).

Leadership is a process that directs a group of individuals and employees to the determined goal with high efficiency and commitment, and the role of the leader in bringing the team to the goal is decisive (Jaques *et al.*, 1994). Leadership is defined as a guide in the process of creating the corporate vision, developing strategies, implementing ideas, and achieving the set goals. In this context, the leader must have the ability to influence and inspire those around him towards the goal to be achieved (Bennis, 1989).

NLP techniques can be used in leadership evaluation through methods such as computer-aided text analysis and supervised and unsupervised machine learning. These approaches are designed to facilitate the analysis of substantial quantities of text data relating to leadership behaviour, communication style, and decision-making. For instance, leaders' written or spoken communication can be evaluated to gain insights into their effectiveness and areas for improvement (Tonidandel and Albritton, 2023). AI-powered tools can provide leaders with instant analytics and recommendations based on their interactions and behaviours, helping them improve their communication skills, decision-making processes, and overall effectiveness (Rajendran and Chitrarasu, 2024).

Digital technologies have permanent effects on human behaviour, business style, and attitudes as they change the dynamics of the working environment as well as the social environment. New business models have emerged and can be implemented thanks to the developing technology. Thanks to remote access and operation, which also prevents resource wastage, interruptions can also be prevented. In addition, thanks to these systems, it is possible to carry out activities in dangerous situations without entering the hazardous environment. Of course, for a limited number of jobs or activities, there are industrial areas where remote access or full capacity systems are operated and controlled, that is, fully autonomous systems. Again, the question of whether there will be unmanned aerial vehicles and passenger aircraft in aviation is being investigated. However, in consumer research, it is concluded that unmanned passenger aircraft will not be preferred yet. The most critical determinant in all systems in which people are involved is still "trust". Without trust, it is not possible to manage society and social events as desired. As all developing technologies, including NLP, progress evolutionarily and as new versions emerge, their effects on all areas, including leadership, management, and strategy, will increase and differentiate.

We have been talking about the human–machine interface in ergonomics/human factors for many years, but today we are talking about the human digital environment and the human–AI interface. Approaches to leadership, whether self-contained or distributed, all now have a digital side as well. The rapid advancements in technology and AI we are witnessing have made digital literacy one of the rules of existence in the professional business world. Therefore, with the contingency approach, every leader today is a digital leader. The way for companies to achieve their resilience and goals is through their digital competencies (Helmold *et al.*, 2024).

The NLP is not merely a technological marvel; it's a bridge between our human world and the silicon one. As research and development push the boundaries of what's possible, we can expect even more sophisticated language models, ethical considerations tackled head-on, and a future where our interactions with technology are not only efficient but also deeply human. The revolution of communication has begun, and NLP is leading the charge (Medium, 2024).

NLP has the potential to offer significant opportunities to managers in strategic decision-making processes in civil aviation, automotive, and service sectors. NLP technologies analyse large data sets to obtain meaningful insights and make operational processes more efficient. In the aviation sector, NLP has the potential to improve customer experience, increase operational efficiency, and improve security. In airlines and airports, customer service processes can be automated through NLP-supported virtual assistants to provide swift and effective service. The evaluation of passenger feedback through sentiment analysis facilitates data-driven decisions, enhancing service quality. Multilingual communication support is a critical element, increasing passenger satisfaction for aviation companies operating on a global scale. In ground services and terminal companies, NLP-supported routing systems can improve the travel experience by optimising passenger flow. NLP-based solutions in baggage tracking and lost property management can minimise operational disruptions by providing rapid responses to passenger requests.

In flight operations, NLP can contribute to increased security by analysing communication between air traffic controllers and pilots, and automatic conversion of voice commands to text allows flight records to be created quickly and without errors. In maintenance, repair and revision processes, automatic summarisation of technical documents with NLP can accelerate fault detection and repair processes by providing maintenance teams with faster access to critical information. As part of preventive maintenance strategies, NLP analyses can be performed on maintenance records to predict potential technical failures and ensure operational continuity.

Communication is central to modern life and business, and in the realm of social media, NLP offers powerful new possibilities. TweetNLP is an advanced NLP platform tailored specifically for Twitter (now known as X), supporting tasks such as sentiment analysis, hate speech detection, irony recognition, emoji prediction, named entity recognition, and topic classification. Built on transformer-based models such as RoBERTa and XLM-R, pretrained on Twitter-specific data, it effectively addresses the challenges of noisy, multilingual, and platform-specific content. The platform includes efficient models such as TimeLMs and XLM-T, designed to run on standard hardware, and provides accessible tools such as a Python library, online demos, and tutorials for users of all levels. With regularly updated models and plans to expand to platforms like Reddit and Instagram, TweetNLP aims to make cutting-edge NLP tools more widely available for social media analysis (Camacho-Collados *et al.*, 2022).

In the automotive industry, NLP-supported smart assistants enable drivers to manage in-car systems with voice commands, thereby enhancing the user experience. In the domain of autonomous vehicle technologies, NLP-based natural language understanding systems can optimise vehicle control processes by facilitating more effective interaction with drivers and passengers. The analysis of customer feedback using NLP provides a strategic advantage in the identification of consumer trends and the data-driven development of products.

In the service sector, the integration of NLP-based chatbots within call centres has the potential to enhance operational efficiency by facilitating swift and precise responses to customer queries. Furthermore, the analysis of employee feedback using NLP techniques enables more informed decisions to be made in HR processes. In training processes, NLP-supported systems can personalise content for employees, thereby optimising learning processes.

5.4 Insights into Optimisation Management and Strategy: Leader Side Effects of NLP

NLP stands out as a technology that can support the optimisation of decision-making processes for leaders and managers. How can it achieve this optimisation? Thanks to their fast and accurate data analysis, predictive modelling, and automation capabilities, NLP systems have the potential to support managers to make faster, more accurate, and strategic decisions. But of course, at this point, the question comes to mind as to whether these NLP outputs should be used directly, that is, whether they should be accepted as they are produced without questioning the NLP results. The solutions offered by NLP increase the effectiveness of decision-making mechanisms in terms of speed, diversity, and versatility, and they maximise organisational efficiency.

Big Data analytics plays a critical role in making sense of the dense and unstructured data that managers often face. Big Data sources such as e-mails, reports, news, social media posts, customer comments, and market analysis can be quickly analysed and transformed into meaningful information through NLP. In this way, managers can have access to accurate and relevant information while saving time in the decision-making process.

Predictive analytics is another important contributor area of NLP, enabling the prediction of future trends by examining past data. Marketing managers can shape their strategies by analysing customer feedback, while decision-makers in the financial sector can optimise their investment planning by examining market data. In this way, companies can act more consciously to gain a competitive advantage and minimise risks.

NLP-based sentiment and perception analysis makes a huge difference in understanding employee and customer satisfaction. By analysing data obtained through social media, surveys, and feedback, businesses can take proactive measures to increase employee engagement and detect customer dissatisfaction early. This is a critical element for organisations to maintain their brand reputation and achieve long-term success.

NLP is also used effectively in the fields of risk management and legal compliance. NLP provides great convenience to decision-makers in various areas, such as detecting fraud in the financial sector, predicting possible risks by analysing critical items in legal texts, and increasing companies' compliance with regulations. Thanks to this technology, potential hazards can be identified early and strategic solutions can be developed.

NLP is considered worthy of inclusion in the process in terms of a lean management approach. However, this does not eliminate the environmental, economic, and social impacts of technological production and development studies.

In addition, the automation capabilities of NLP make routine decision-making processes more efficient. Thanks to NLP-based systems, HR managers can quickly analyse thousands of resumes and identify the most suitable candidates, while supply chain managers can predict possible disruptions by scanning news sources. This automation process allows managers to focus on strategic issues, saving time and money.

The potential implications of NLP for risk leaders and top-level executives are significant, as advancements in NLP have the capacity to redefine human interaction with technology. The future holds immense possibilities, ranging from hyper-personalised user experiences and smarter business insights to breakthroughs in healthcare and other fields. By embracing the cutting-edge innovations and transformative applications of NLP, businesses and individuals alike can unlock unprecedented opportunities. The future of NLP lies not only in its capacity to process language but also in its ability to understand and leverage the profound knowledge embedded within it (Pragada, 2024).

NLP-based data analytics solutions for managers facilitate objective and rapid decision-making in strategic processes by enabling the interpretation of large volumes of data. The automation and analysis capabilities of NLP enhance operational efficiency while also fostering customer and employee satisfaction. In this context, the integration of NLP into strategic decision-making mechanisms should be regarded as a significant opportunity for providing competitive advantage and developing innovative business models.

The need for self-leadership is increasing in a professional and social environment that is digitalising and evolving under the influence of AI. Self-leadership and other leadership approaches are the only ways to be resilient and manage risk in such a complex, accelerated environment where decision-making becomes difficult. In an environment that is becoming increasingly automated, the sharing of leadership and the multifaceted nature of people appear to be other necessary elements (Helmold *et al.*, 2024).

5.5 Unlocking the Language of the Future: Advances in NLP

Advances in NLP are bringing us to the cutting edge of a new era in AI and promise to redefine how we interact with technology. The future holds immense possibilities, including hyper-personalised user experiences, smarter business insights, and breakthroughs in healthcare and beyond. By embracing the cutting-edge innovations and transformative applications of NLP, businesses and individuals alike can unlock unprecedented opportunities. The journey ahead is not just about processing language; it is about understanding and harnessing the deep knowledge embedded within it (Pragada, 2024).

From a management and strategy perspective, digital transformation and NLP assist leaders and managers in optimising the decision-making processes of managerial functions, thereby facilitating faster, more accurate, and data-driven decision-making. Nevertheless, it is imperative to question the reliance on data alone in decision-making, as leaders must also exercise innovative and strategic interpretations and predictions in that process. Conversely, a decision-making process that is solely reliant on data may result in a homogeneous decision landscape, potentially leading to diminished competitiveness. In critical domains such as the analysis of voluminous and heterogeneous data, the construction of predictive forecasts based on analytical results, sentiment analysis, risk management, and operational efficiency, NLP holds the potential to enhance organisational effectiveness and facilitate the timely and precise allocation of resources. However, erroneous insights may emerge if the systems are not designed effectively or if the results are not interpreted correctly by the leaders. Consequently, a collaborative design of the human–machine interface and an approach that prioritises human decision-making appear to be more suitable.

Bibliography

Airbus (2024) 'How Airbus Uses Generative Artificial Intelligence to Reinvent Itself', 28 May. Available at: https://www.airbus.com/en/newsroom/stories/2024-05-how-airbus-uses-generative-artificial-intelligence-to-reinvent-itself (Accessed: 28 May 2024).

Alsarrani, W.I. and Jusoh, A. (2019) 'A Review on the Characteristics of a Good Quality Leadership Style', *Humanities & Social Sciences Reviews*, 7(5), pp. 1152–1159. doi: 10.18510/hssr.2019.75152.

AWS Amazon (2025) 'What are NLP Use Cases for Business?', 11 May. Available at: https://aws.amazon.com/what-is/nlp/ (Accessed: 11 May 2025).

BBC News (2025) 'Plane Crash in the US: Trump Blames Helicopter Pilot', 19 February. Available at: https://www.bbc.com/turkce/articles/c203edxj2zwo (Accessed: 19 February 2025).

Bennis, W. (1989) *On Becoming A Leader*. New York: Basic Books.

Camacho-Collados, J. *et al.* (2022) 'TweetNLP: Cutting-Edge Natural Language Processing for Social Media', *Proceedings of the 2022 Conference on Empirical Methods in Natural Language Processing: System Demonstrations*. Abu Dhabi: Association for Computational Linguistics, pp. 38–49. doi: 10.18653/v1/2022.emnlp-demos.5.

Ghasabeh, M.S., Soosay, C. and Reaiche, C. (2015) 'The Emerging Role of Transformational Leadership', *The Journal of Developing Areas*, 49(6), pp. 459–467. doi: 10.1353/jda.2015.0090.

Hirschberg, J. and Manning, C.D. (2015) 'Advances in Natural Language Processing', *Science*, 349(6245), pp. 261–266. doi: 10.1126/science.aaa8685.

Jaques, E., Clement, S.D. and Lessem, R. (1994) *Executive Leadership: A Practical Guide to Managing Complexity*. Oxford: Blackwell.

Medium (2024) 'Revolutionizing Communication: The Power of NLP and its Cutting-Edge Future'. Available at: https://medium.com/@jibintcr/revolutionizing-communication-the-power-of-nlp-and-its-cutting-edge-future-13f517007f57 (Accessed: 31 December 2024).

Mersha, B.W., Jansen, D.N. and Ma, H. (2023) 'Angle of Attack Prediction Using Recurrent Neural Networks in Flight Conditions with Faulty Sensors in the

Case of F-16 Fighter Jet', *Complex & Intelligent Systems*, 9, pp. 2599–2611. doi: 10.1007/s40747-021-00612-6.

Napkin AI. Available at: https://app.napkin.ai/page/CgoiCHByb2Qtb25lEiwKBFBhZ2UaJDY1MmY2Y2UzLWNiYTktNDc3ZS04MzE1LWVmY2Y1MmI2NzJiMg.

Paul, S., Purkaystha, B.S. and Das, P. (2018) 'NLP Tools Used in Civil Aviation: A Survey', *International Journal of Advanced Research in Computer Science*, 9(2), pp. 109–114. doi: 10.26483/ijarcs.v9i2.5559.

Pragada, S.V. (2024) 'The Cutting Edge of Natural Language Processing (NLP): Innovations and Transformative Applications', *LinkedIn*, January. Available at: https://www.linkedin.com/pulse/cutting-edge-natural-language-processing-nlp-sasi-varun-pragada-u5cec/ (Accessed: 9 January 2024).

Rajendran, R.K., T., M.P. and Chitrarasu, K. (2024) 'Natural Language Processing (NLP) in Chatbot Design' in *Design and Development of Emerging Chatbot Technology*. Hershey: IGI Global, pp. 102–113.

Santos, L. *et al.* (2025) 'The Role of Artificial Intelligence in Optometric Diagnostics and Research: Deep Learning and Time-Series Forecasting Applications', *Technologies*, 13(2), 77. doi: 10.3390/technologies13020077.

Spiceworks (2025) '3 Ways Natural Language Processing (NLP) Will Transform HR in 2019', 10 May. Available at: https://www.spiceworks.com/hr/hr-strategy/articles/3-ways-natural-language-processing-nlp-will-transform-hr-in-2019/ (Accessed: 10 May 2025).

Tanguy, L. *et al.* (2016) 'Natural Language Processing for Aviation Safety Reports: From Classification to Interactive Analysis', *Computers in Industry*, 78, pp. 80–95. doi: 10.1016/j.compind.2015.09.005.

Tonidandel, S. and Albritton, B.H. (2023) 'Transforming Leadership Assessment Using Natural Language Processing' in Kantrowitz, T.M. (ed.) *Talent Assessment: Embracing Innovation and Mitigating Risk in the Digital Age*. Oxford: Oxford University Press, pp. 149–168.

Yamin, T. and Murwaningsari, E. (2023) 'Exploring the Interplay Between Digital Technology, Transformational Leadership and Agility for Enhancing Organisational Performance', *Business Ethics and Leadership*, 7(4), pp. 73–88.

Chapter 6

Types and Categories of Realities

6.1 Extended Realities

Realities are the sum or aggregate of everything in existence, as opposed to that which is only imaginary or nonexistent. Different cultures and academic disciplines conceptualise it in various ways.

Extended reality (XR) is an umbrella term to refer to augmented reality (AR), mixed reality (MR), and virtual reality (VR). The technology is intended to combine or mirror the physical world with a "digital twin world" able to interact with it, giving users an immersive experience by being in a virtual or augmented environment. The fields of VR and AR are rapidly growing and being applied across a wide range of areas, including entertainment, cinema, marketing, real estate, training, education, maintenancc, and rcmotc work. XR has thc ability to bc uscd for collaborativc working, training, educational purposes, therapeutic treatments, and data exploration and analysis.

XR works by using visual data acquisition that is either accessed locally or shared and transferred over a network to the human senses. By enabling real-time responses in a virtual stimulus, these devices create customised experiences. Advancing in 5G and edge computing – a type of computing that is done "at or near the source of data" – could aid in data rates, increase user capacity, and reduce latency. These applications will likely expand XR into the future (Lang and Müller, 2020).

Realities enable multidimensional content viewing and collaboration throughout the value chain through cloud-based systems and in real-scale environments, resulting in improved processes and higher levels of efficiency. Realities help to eliminate waste in all stages of a product lifecycle and therefore offer an ideal and integrative lean management approach that combines project management philosophies, systems, technologies, and tools (Alizadehsalehi and Hadavi, 2023). XR advances the potential of AR, VR, and MR, merging our real and virtual worlds to create new environments and visualisations where physical and digital objects co-exist, interact, and communicate (Marr, 2021). Instead of removing users completely from the real world or simply layering flat content on top of their immediate view, MR adds intelligence, even personality, to digital content relative to the world around them (Helmold, 2023). As part of the technological evolution of how we engage with the digital world, in both our personal and work lives, we are smashing through the barriers that interfere with our ability to make smart decisions quickly, absorb, retain, and process critical information, visualise possible scenarios before acting, or share knowledge and tasks. The XR trend is ushering in a new world of simulated experiences grounded in the ways business gets done and how customers actually use products. AR or XR enables a gradual, full or partial, virtual representation of reality, which can either closely resemble reality or completely deviate from it. The boundaries between reality and the virtual world are becoming increasingly blurred thanks to new hardware and computing power, as well as high-performance networks with high bandwidth, opening up new, breathtaking impressions that until recently could only have been imagined in science fiction films. AR is also a key technology for what is known as the Metaverse, the third major generation of the internet, often referred to as Web 3.0. This follows on from documents and their linking, based on the first and second generations shaped by social media. AR systems have been researched and developed for quite some time. The first systems for computer games were already commercially available in the 1990s. However, only in the past 10 years have systems been developed that enable good performance while remaining cost-effective. The simplest systems consist of cardboard constructions into which the smartphone can be installed.

Figure 6.1 gives an overview of XR (virtual, mixed, and augmented realities), which are detailed in the following sections.

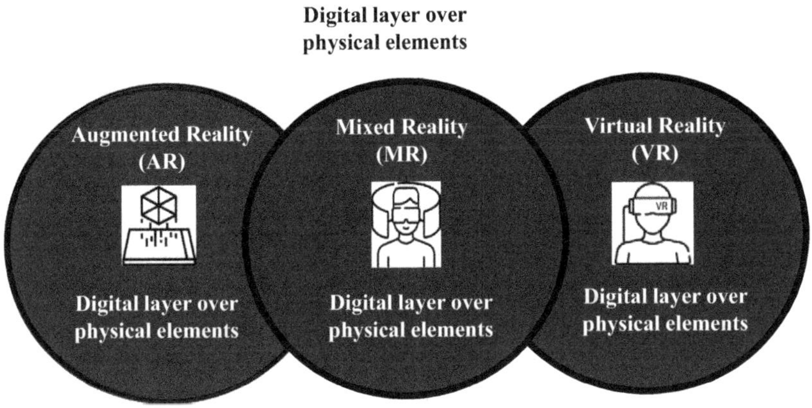

Figure 6.1. Extended realities (XRs).

6.2 Creation of XR

The realities mentioned above are generated by suitable hardware. First of all, these include glasses, which are also referred to as head-mounted displays (HMDs). There are three categories: connected (tethered), wireless (mobile), and fully autonomous (standalone) devices.

- **Tethered:** These glasses must be tethered and connected to a powerful computer system.
- **Wireless (mobile):** This solution consists of an attachment in which smartphones approved for this purpose can be used. A cable connection is no longer required, which significantly increases the range of use and enables new applications.
- **Autonomous (standalone):** Devices that do not require a connection to a computer or the use of a smartphone are referred to as autonomous or standalone. They offer the greatest possible flexibility and the implementation of demanding applications.

Regardless of this category, the following parameters are important when choosing a suitable solution:

- display resolution and type;
- resolution and visualisation capability;

- connectivity via USB, Bluetooth, etc.;
- position and location tracking options;
- field of views and simulations;
- configurability with other devices.

In standalone systems, the processor (e.g., Intel, Snapdragon, or Nvidia) and the platform (Windows MR or Google Daydream) play a particularly important role.

Of course, different systems are possible for VR and AR applications. In the field of VR, HTC Vive and Oculus Rift are very popular systems. Microsoft HoloLens is a leading system for AR applications. For the realisation of AR on mobile phones, Apple's ARKit and Google's ARCore are very widespread and powerful systems.

In order to bring the virtual world closer to reality, HMDs can be supplemented with wearable elements such as tactile gloves, suits, or vests. The impression in the virtual world is thereby again clearly strengthened. Haptic gloves enable the position of fingers and hands to be recorded, thereby creating another interface for control between humans and machines. Such gloves also enable feedback, such as vibrations, impressions of force, or even the feeling of surface textures. Holding and interacting with objects can be simulated using controllers.

Only the combination of these hardware elements with suitable software then enables the creation of a suitable XR system.

The recent trends for market growth from US$12 billion in 2020 to over US$72 billion in 2024 show the great potential of this technology. Of course, this development is supported by the games market, but these systems are also increasingly found in production environments. However, some prerequisites must be established for their smooth use. This includes, for example, sufficient lighting conditions or the detection and removal of highly reflective parts (Helmold, 2024). The global XR market size is projected to grow from US$131.54 billion in 2023 to US$1,134.79 billion by 2030, at a CAGR of 36.0 percent (Helmold, 2024).

6.2.1 Augmented reality

Creative collaboration demands separation from the context of the physical real world. As a sub-mode of MR based on the definitions by Milgram and Colquhoun (1999), AR can add virtual elements into the physical real world, which can be associated with tangible interfaces and provide great benefits in architecture and urban design (Seichter and Schnabel 2005). Though real elements in AR could potentially place certain constraints on the shared imagination, major advantages of AR include ease of collaboration, intuitive interaction, integration of digital information, and mobile computing. AR allows a user to work in a real-world environment while visually receiving additional computer-generated or modelled information to support the task at hand. In the past, AR environments have been applied primarily in scientific visualisation and gaming entertainment. In recent years, it has been explored for goal-oriented human activities such as surgery, training, and collaborative work.

6.2.2 Mixed reality

MR (Milgram and Colquhoun, 1999), a more expansive form of VR, describes a concept where computer-generated content and the real world intermingle. It offers new potential for interaction between design information and collaborators for the entire lifecycle of the engineered facility. Depending on the method of augmentation, MR involves two major modes: AR and augmented virtuality, which will be discussed in the two sections that follow. MR has been applied to enhance design practices. For instance, Wang *et al.* (2003) developed an intuitive mixed environment called Mixed Reality-based Collaborative Virtual Environment (MRCVE) to support collaboration and design spatial comprehension in collaborative design review sessions for mechanical contracting. The environment supported both face-to-face and distributed communication. A slightly different interpretation of MR is presented in Schnabel *et al.* (2004), where a cyclical design process that interchanges between real and virtual realms in order to develop and communicate

MR enables interaction with and manipulation of digital or virtual and real objects in real time. It is sometimes also referred to as hybrid reality and enables interaction with virtual objects. MR requires a suitable headset (e.g., Microsoft HoloLens) and significantly more computing power than VR or AR. Virtual objects can then be placed in space, rotated, and further interacted with. MR applications can also be implemented on smartphones or tablets, although the boundaries between AR

Mixed Reality

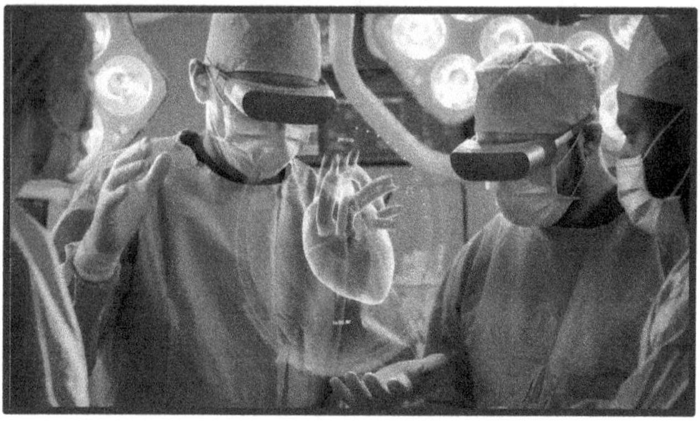

Figure 6.2. Mixed realities (MRs) in healthcare.

and MR are becoming increasingly blurred. One example is Google 3D Experience, which allows virtual objects such as animals to be placed and scaled in real space. This can be used for training, for example. It is also possible to get an idea of how certain objects, such as furniture, fit into a room.

Figure 6.2 shows an example of MR in healthcare (Helmold *et al.*, 2023).

6.2.3 Virtual reality

VR visualisation technology, which creates a total virtual environment (VE), was originally embraced by architecture visionaries for design concept presentations during the recent decades. As computing has advanced, they have supported more sophisticated graphics capabilities (Hendrickson and Rehak, 1993). VR research attracts a considerably wide range of interest in diverse fields, with the typical applications being web-based design review, computer-supported cooperative work (CSCW), and 3D-type thinking using desktop PCs and headset-based VEs. In 2001, for the first time, immersive VEs were successfully employed to create and communicate architectural design in a broader context (Schnabel, 2002). In VEs, subtle human-to-human interaction is locked into these environments, which often provide only limited support for the physical expression of emotion and personality. The notions of VRs are relatively familiar compared with other concepts and have been widely applied in many areas of design and research.

VR is a simulated experience that employs pose tracking and 3D near-eye displays to give the user an immersive feel of a virtual world. VR is the use of computer technology to create a simulated environment which can be explored in 360 degrees. Unlike traditional interfaces, VR places the user inside the virtual environment to give an immersive experience. VR enables users to explore and interact with a virtual surrounding in a way that approximates reality as it is perceived through the users' senses. The environment is created with computer hardware and software, although users might also need to wear devices such as helmets or goggles to interact with it. Users can thus immerse themselves more deeply in a VR environment.

The VR industry still has far to go before realising its vision of a totally immersive environment that enables users to engage multiple sensations in a way that approximates reality. However, the technology has come a long way in providing realistic sensory engagement and shows promise for business use in a number of industries.

VR systems can vary significantly from one another, depending on their purpose and the technology used, although they generally fall into one of the following three categories:

- **Non-immersive:** This type of VR typically refers to a 3D simulated environment that is accessed through a computer screen. The environment could also generate sound, depending on the program. The user has some control over the virtual environment using a keyboard, mouse, or other devices, but the environment does not directly interact with the user. A video game is a good example of non-immersive VR, as is a website that enables a user to design a room's decor.
- **Semi-immersive:** This type of VR offers a partial virtual experience that is accessed through a computer screen or some type of glasses or a headset. It focuses primarily on the visual 3D aspect of VR and does not incorporate physical movement in the way that full immersion does. A common example of semi-immersive VR is the flight simulator, which is used by airlines and militaries to train their pilots.
- **Fully immersive:** This type of VR delivers the greatest level of VR, completely immersing the user in the simulated 3D world. It incorporates sight, sound, and, in some cases, touch. There have even been

Figure 6.3. Virtual reality (VR) in engine production.

some experiments with the addition of smell. Users wear special equipment such as helmets, goggles, or gloves and are able to fully interact with the environment. The environment might also incorporate such equipment as treadmills or stationary bicycles to provide users with the experience of moving through the 3D space. Fully immersive VR technology is a field still in its infancy, but it has made important inroads into the gaming industry and, to some extent, the healthcare industry, and it is generating a great deal of interest in others.

Figure 6.3 illustrates an example of VR in engine production (Helmold *et al.*, 2023) at BMW. VR supports the production process by preventing the occurrence of quality errors. VR also helps operators with tool usage, operational sequences, and quality assurance.

6.3 Extended View in the World of Realities

Apart from the three realities (augmented, virtual, and mixed), there are classifications which extend the types of realities to a total of six (see Figure 6.4). The additional realities are:

- mediated reality (MedR),
- real reality (RR),
- amplified reality (AmR).

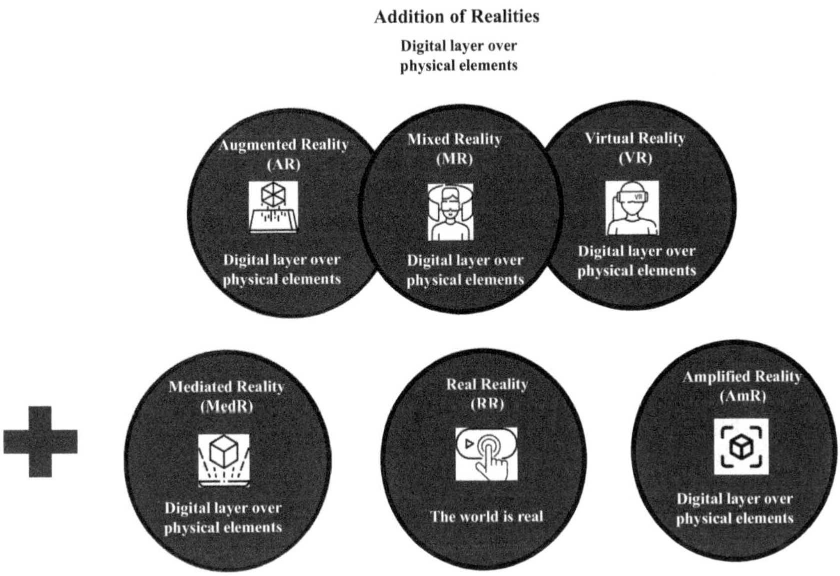

Figure 6.4. Addition of mediated, real, and amplified realities.

6.4 Mediated Realities

MedR refers to a more general framework for artificial modification of human perception by way of devices that augment (Starner *et al.*, 1997), deliberately diminish, and more generally alter sensory input. This is typically achieved by altering a video input stream light that would normally directly reach the user's eyes and computationally filtering it into a more useful form. MedR is a broad term encompassing technologies that modify human perception through computer processing, effectively creating an adjusted, filtered, or amended version of reality.

6.5 Real Realities

RR describes the world as it is. The real world utilises machines, computers, and equipment, but it is not a virtual, augmented, or online world. RR shows the world and objects in the world as they are.

6.6 Amplified Realities

To amplify reality is to enhance the perceptible properties of a physical object by means of using embedded computational resources. AmR (Falk *et al.*, 1999) was originally introduced as a concept complementary to that of AR. Compared with AR, which enhances impressions of real objects, AmR amplifies the expressions of objects in the real world and emphasises the importance of the shared experiences that result from publicly available properties of objects. AR is about how the user perceives reality, while AmR is based on how the perceived might control how information is made available. AmR has been applied in the projects of the Lovegety (Iwatani, 1998), the Hummingbird (Holmquist *et al.*, 1999), and the BubbleBadge (Falk and Björk, 1997). An amplified object is self-contained in regard to its properties. In practice, this means that the properties are embedded parts of the object. In contrast, AR superimposes virtual properties onto an object, which in effect does not change the actual object but rather how we perceive or experience it.

Bibliography

Alizadehsalehi, S. and Hadavi, A. (2023) 'Synergies of Lean, BIM, and Extended Reality (LBX) for Project Delivery Management', *Sustainability*, 15(6), 4969.

Fritz, J. (2022) *Datenbasierte Optimierung des Business Management Systems — Geschäftsprozesse verbessern mit Data Analytics, Industrie 4.0, KI, Chatbots und Co.* München: Hanser Verlag.

Helmold, M. (2024) *Erfolgreiche Transformation zum digitalen Champion: Wettbewerbsvorteile durch Digitalisierung und Künstliche Intelligenz*. Wiesbaden: Springer.

Helmold, M. *et al.* (2024) *The Digital Champion: Best Practices and Insights for the Successful Transformation of Enterprises and Companies*. Singapore: World Scientific.

Lang, M. and Müller, M. (2020) *Von Augmented Reality bis KI – Die wichtigsten IT-Themen, die Sie für Ihr Unternehmen kennen müssen*. München: Hanser Verlag.

Marr, B. (2021) *Extended Reality in Practice: 100+ Amazing Ways Virtual, Augmented and Mixed Reality Are Changing Business and Society*. Chennai: Wiley.

PTC (2019) 'Volvo Case Study–Volvo Group Delivers Digital Thread Through the Lens of Augmented Reality'. Available at: https://www.ptc.com/de/case-studies/volvo-group-digital-thread (Accessed: 14 May 2022).

Richter, B. and Vogel, U. (2021) 'Technisch reif für Assisted Reality', *Elektronik*, 07, pp. 14–19.

Schnabel, M.A. *et al.* (2007) 'From Virtuality to Reality and Back', *International Association of Societies of Design Research (IASDR)*. Hong Kong, November 2007.

Volvo (2022) *Leading the Transformation–Volvo Group Annual and Sustainability Report 2021*. Available at: https://www.volvogroup.com/content/dam/volvo-group/markets/master/investors/reports-and-presentations/annual-reports/annual-and-sustainability-report-2021.pdf (Accessed: 14 May 2022).

Welling, D. (2019) 'Control 2019 Quality Checks with Augmented Reality', *Pressemitteilung Fraunhofer IGD*.

Chapter 7

Cryptocurrencies and Blockchain

7.1 The Hype of Bitcoin

The 2008 global financial crisis exposed systemic weaknesses in the financial system, including excessive risk-taking by commercial banks, a lack of transparency, and regulator failures in preventing a major economic collapse. In this climate of mistrust towards traditional financial institutions and centralised regulatory authorities, a whitepaper titled "Bitcoin: A Peer-to-Peer Electronic Cash System" was published under the pseudonym Satoshi Nakamoto, introducing Bitcoin – a novel form of digital currency – and its underlying technology, blockchain, to the world (Nakamoto, 2008), paving the way for a new era of financial innovation (Deloitte, n.d.).

Traditionally, a financial transaction such as a money transfer depends on the services of trusted third parties (intermediaries) such as commercial banks, investment banks, or pension funds to process payments. That means the intermediaries help move funds from the originator (sender of the fund) to the beneficiary (recipient of the fund). In an electronic money transfer, the process generally involves a series of electronic messages sent between the banks of the originator and the beneficiary. These messages direct each bank to make the necessary debit and credit accounting entries to complete the transaction without necessarily transporting the money bills (see Figure 7.1).

In contrast to traditional financial transactions, Bitcoin operates in a decentralised user network without any intermediary. In other words, Bitcoin transactions can take place directly between users in the network

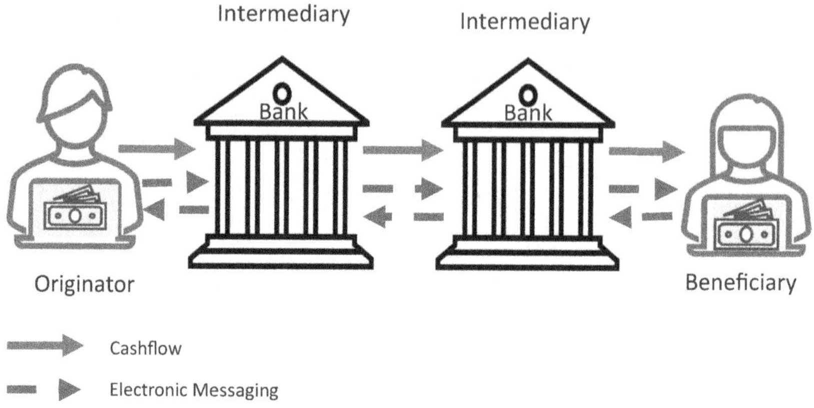

Figure 7.1. A traditional financial transaction.

("peer-to-peer"). The system issues tokens intended as a means of payment based on a decentralised and immutable digital ledger. The transactions are verified and recorded through a distributed consensus mechanism known as Distributed Ledger Technology (DLT). At its core, DLT uses cryptographic methods to ensure the authenticity and integrity of the transactions, making the entries in the decentralised ledger verifiable and irreversible, minimising the risk of fraud, censorship, and manipulation. Bitcoin transactions are more cost-efficient without the costs associated with traditional intermediaries (Nakamoto, 2008) (see Figure 7.2).

The Bitcoin network was launched on 3 January 2009, marked by the mining of the Genesis Block ("Block 0"). This block contained the text "The Times 03/Jan/2009 Chancellor on brink of second bailout for banks", a reference to a headline in *The Times* newspaper that suggested the need for an alternative financial system. This message highlighted the financial instability and failed government interventions during the 2008 financial crisis, emphasising Bitcoin's potential as a decentralised alternative to traditional banking systems (European Commission, 2019).

The total supply of Bitcoin is limited to 21 million coins, a feature embedded in its protocol to create scarcity and prevent inflation (Nakamoto, 2008). Bitcoin's rapid price surges have garnered significant media attention, driving increased interest from investors and the public alike. Bitcoin's success paved the way for numerous alternative cryptocurrencies, or "altcoins", such as Ethereum (ETH), Tether (USDT), and Dogecoin (DOGE). Altcoins share Bitcoin's decentralised, peer-to-peer

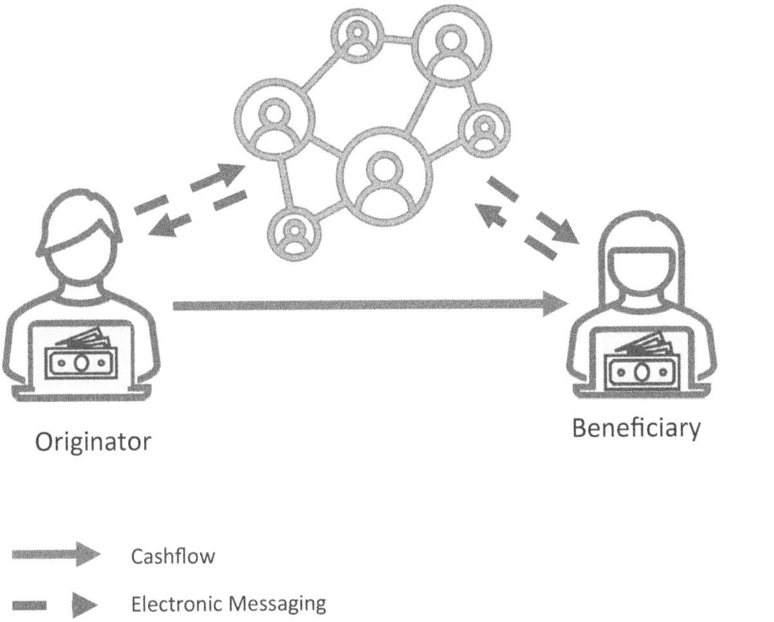

Figure 7.2. A cryptocurrency financial transaction.

principles but operate on their own blockchains with unique rules (European Commission, 2019).

Cryptocurrencies have grown significantly since Bitcoin's creation in 2009 and are now worth over US$1 trillion collectively. Despite their extremely high volatility and regulatory challenges, cryptocurrencies have gained traction as investment vehicles and mediums of exchange (U.S. Council on Foreign Relations, 2024).

7.2 Distributed Ledger Technology, Blockchains, and Cryptocurrencies

DLT refers to a digital system for recording transactions of assets in which the transactions and their details are recorded in multiple places at the same time. In contrast to traditional databases that have a central data store, distributed ledgers do not have a central administrator. DLT allows for the recording, sharing, and synchronisation of transactions and data across a network of participants, using digital wallets, accessible via computers, mobile devices, or internet portals. All network participants maintain their own identical copy of the ledger, ensuring transparency and reducing the risk of data tampering or loss (Hedera, n.d.). To validate transactions and ensure all copies of the ledger are identical, the system uses consensus algorithms. Common validation mechanisms include Proof of Work (PoW), Proof of Stake (PoS), and Practical Byzantine Fault Tolerance (PBFT). These algorithms ensure that all nodes agree on the state of the ledger (Sunyaev, 2024)

Blockchain technology is a specific type of DLT. It organises data into blocks, which are linked together in a chronological chain. Each block contains a batch of transactions, a timestamp, and a cryptographic hash of the previous block, ensuring data integrity. Unlike traditional centralised databases, blockchains operate on a peer-to-peer network, where each participant ("node") maintains a copy of the ledger. This eliminates the need for a central authority and enhances security (Hedera, n.d.).

Blockchain technology, based on a decentralised and immutable distributed ledger, has laid the foundation for cryptocurrencies such as Bitcoin. Cryptocurrencies are digital or virtual currencies that use cryptographic techniques to secure transactions. At its core, a cryptocurrency system is designed for the issuance of digital tokens intended for use as a medium of exchange within a digital ledger, employing cryptography to ensure security and trust in transactions. Cryptocurrencies can be sent and received anywhere in the world, making them accessible to anyone with an internet connection. At present, the most influential cryptocurrencies include the following (Coinmarket, 2025):

- **Bitcoin (BTC):** The first and most well-known cryptocurrency, created by an anonymous person or group known as Satoshi Nakamoto in 2009.
- **Ethereum (ETH):** Known for its smart contract functionality, Ethereum allows developers to build decentralised applications (dApps) on its blockchain.
- **Ripple (XRP):** Focuses on enabling real-time, cross-border payment systems for financial institutions.
- **Litecoin (LTC):** Created as a "lighter" version of Bitcoin, offering faster transaction times and a different hashing algorithm.

In the Bitcoin network, for example, users can send and receive bitcoins using a pair of cryptographic keys: a public key (Bitcoin user ID) for receiving payments and a private key for authorising payments. Once

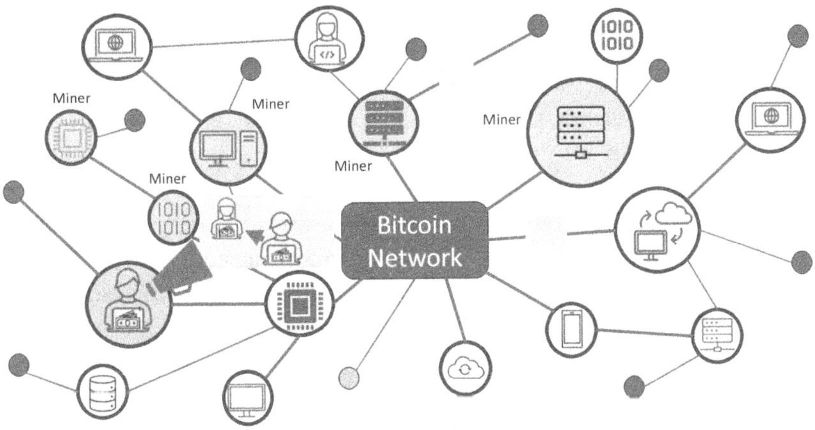

Figure 7.3. The broadcast of a Bitcoin transaction.

initiated, each transaction must be broadcast by the sender (originator) to the network, including the amount and the public keys of the sender and the receiver (see Figure 7.3). The message bears a digital signature of the sender created with their private key, ensuring that only the owner of the Bitcoin account can authorise the transaction (Nakamoto, 2008).

Bitcoin transactions are verified by network nodes (individuals or entities known as "miners") using cryptographic algorithms and recorded in a public, decentralised ledger called the blockchain. Unconfirmed transactions are initially collected and verified by the miners, ensuring that the transactions comply with the transaction rules on the Bitcoin network. The key transaction rules include the following (Nakamoto, 2008):

- The sender possesses sufficient Bitcoins for the transaction.
- The inputs have not been previously spent (to prevent double-spending).
- The total value of the inputs is higher than the total value of the outputs. The difference between total inputs and outputs is considered a transaction fee for the minor.

For the purpose of efficient validation, transactions are grouped into a data block by the miners. To validate a new block, Bitcoin miners compete to find a unique string of numbers and letters that solves a complex cryptographic puzzle. This process usually requires significant computational capacity and is known as PoW. The first miner to solve the puzzle broadcasts the solution to the network. The new block is then added to the existing chain of blocks ("blockchain") in chronological order. Each block contains the verified transactions and is cryptographically linked to the previous one, ensuring transparency and integrity of the network. The updated blockchain information is copied to all nodes in the network, making manipulation nearly impossible (Nakamoto, 2008).

In return for their efforts in the validation process, successful miners are rewarded with newly created bitcoins and transaction fees from the transactions included in the block. This reward system incentivises miners to continue their participation in the validation process (Nakamoto, 2008).

Cryptocurrencies operate on two distinct types of blockchains: permissionless and permissioned blockchains. Permissionless blockchains are ideal for public applications where transparency, decentralisation, and security are paramount, such as cryptocurrencies and public dApps. That means anyone can join and participate in the network without needing approval from a central authority. These blockchains are maintained by a distributed network of nodes, ensuring no single entity has control over the entire network. All transactions are publicly recorded on the blockchain, making the data accessible and verifiable by anyone. They use cryptographic algorithms and consensus mechanisms (such as PoW or PoS) to secure the network and validate transactions. Due to their decentralised nature, it is difficult for any single entity to censor or alter transactions on the network (Cointelegraph, 2024; Blcokchain Council, 2024).

Bitcoin's permissionless blockchain allows its users to maintain anonymity, despite the transparency of all transactions, because the identities of the participants are kept pseudonymous. Cryptocurrencies like Bitcoin offer access to financial services for the unbanked populations, enabling individuals without traditional banking systems to participate in the global economy (World Economic Forum, 2021).

There are also permissioned blockchains that are suitable for private or consortium-based applications where privacy, control, and efficiency are important, such as supply chain management, interbank settlements, and enterprise solutions. In other words, only authorised participants can join the permissioned blockchain network, and access is controlled by a central authority or consortium. These blockchains are often maintained by a known group of participants, providing more control over the network. Transactions and data are only accessible to authorised participants, ensuring privacy and confidentiality. Permissioned blockchains can be more efficient and faster since they do not require extensive consensus mechanisms like those in permissionless blockchains. They can be tailored to specific organisational needs, including setting permissions and roles for different participants. For example, the cryptocurrency Ripple is run on a permissioned blockchain (Cointelegraph, 2024; Blcokchain Council, 2024).

7.3 Decentralised Finance (DeFi)

7.3.1 Emergence and principles of DeFi

The popularity of Bitcoin has sparked a new trend in digital finance known as decentralised finance (DeFi). The term "decentralised finance" refers to financial products built on a programmable, permissionless blockchain that operates without traditional intermediaries such as banks, brokers, or payment processors. As a result, DeFi providers offer financial services directly to their users without the need for intermediaries (Ethereum, 2025; Dathe, 2025).

The DeFi ecosystem is characterised by its open, permissionless nature, allowing anyone with an internet connection to lend or borrow funds, speculate on asset prices using derivatives, trade cryptocurrencies, insure against risks, and earn interest on the DeFi platforms, without needing approval from a central authority. This inclusivity expands access to financial services globally, particularly in regions with limited traditional banking infrastructure. Besides, in contrast to traditional banking services, DeFi systems operate around the clock without restrictions on access times (Ethereum, 2025; Dathe, 2025).

Technically, DeFi relies on smart contracts – self-executing contracts with the terms directly written into lines of code – to automate financial transactions securely and transparently. By eliminating intermediaries, DeFi enhances inclusivity, increases efficiency, and reduces transaction costs, challenging the traditional financial systems based on centralised institutions (Dathe, 2025; Gruin and Knaack, 2020).

7.3.2 Components of the DeFi ecosystem

The DeFi ecosystem is a rapidly evolving sector at the intersection of blockchain technology, digital assets, and financial services. DeFi aims to recreate traditional financial instruments in a decentralised architecture, eliminating the need for intermediaries such as banks and brokers (Chainlink, 2024).

The DeFi ecosystem is highly composable, enabling the combination of different financial services to create new products and services. The key components of the DeFi ecosystem include various financial applications and services, such as the following:

- **Stablecoins:** Cryptocurrencies designed to maintain a stable value. Combining blockchain technology's efficiency with price-stabilising algorithms, stablecoins enable instant, low-cost cross-border transactions and are widely used for lending, borrowing, and liquidity provision. The main types of stablecoins include (Coinbase, 2025):
    - **Fiat-pegged stablecoins**, which are backed by reserves of conventional currencies such as the U.S. dollar, ensuring redemption at a fixed rate.
    - **Crypto-backed stablecoins**, which are collateralised by cryptocurrencies (for example, DAI is supported by reserves of Ethereum) to offset volatility. These stablecoins are often over-collateralised to overcome price fluctuations. For example, if someone wants to issue $100 worth of DAI, they might need to deposit US$150 worth of ETH into a smart contract. If ETH's price drops, the system may require additional collateral or liquidate the assets to maintain stability.
    - **Algorithmic stablecoins**, which are built on smart contracts and supply adjustments.
- **Decentralised Exchanges (DEXs)** are platforms that allow users to trade cryptocurrencies directly with one another without relying on a central authority. DEXs use smart contracts to automate and enforce trades according to predefined rules without the need for intermediaries. DEXs do not hold users' funds, reducing the risk of hacks and thefts associated with centralised exchanges. Users maintain control of their private keys and funds. The popular DEXs include the Ethereum-based Uniswap and SushiSwap (OKX, 202).
- **Lending and Borrowing Platforms** such as Aave and Compound enable users to lend their crypto assets to earn interest or borrow assets. Interest rates on DeFi platforms are determined by supply and demand dynamics within the platform. Lenders earn interest on their deposited assets, while borrowers pay interest on the borrowed amount. In addition, borrowers must provide collateral that exceeds the value of the

loan to mitigate the risk of default ("over-collateralisation") (Binance, 2019).

- **DeFi Insurance Platforms**, such as Nexus Mutual, offer protection against various risks, including smart contract vulnerabilities, hacks, and protocol failures. This coverage is crucial for users who want to safeguard their investments in DeFi protocols. Many DeFi insurance platforms are governed by their communities, allowing token holders to vote on key decisions, such as claim approvals and policy changes (CoinGecko, 2024).

While the DeFi ecosystem offers increased accessibility, transparency, and efficiency compared to conventional financial systems, they also face additional challenges such as smart contract vulnerabilities, regulatory uncertainties, and high market volatility.

7.3.2.1 *Central bank digital currencies*

Central bank digital currencies (CBDCs) are digital representations of a country's official fiat currency, issued and regulated by the central bank. Unlike physical cash, CBDCs exist solely in digital form but retain the same legal status and function as traditional money. They serve as a direct liability of the central bank, meaning that they are backed by the government and do not carry counterparty risk, unlike commercial bank deposits (Genc and Takagi, 2024).

Like fiat currencies, CBDCs differ from private cryptocurrencies such as Bitcoin or Ethereum in being centrally controlled and designed to maintain price stability. Since CBDCs are issued and backed by central banks, they are less susceptible to the volatility and speculative risks associated with cryptocurrencies and private digital assets. Additionally, they are envisioned as a secure, efficient, and modern form of money that leverages digital infrastructure to facilitate transactions (Genc and Takagi, 2024; Dathe, 2025).

Like private cryptocurrencies, CBDCs offer alternative access to financial transactions and products through internet connectivity. Thus, they have the potential to provide financial services to disadvantaged population groups, such as the physically challenged or people in unbanked and underbanked regions, over the internet, accessed from their mobile phones (CNBC, 2021; Dathe, 2025).

CBDCs can modernise national payment systems by enabling faster, cheaper, and more secure transactions. By leveraging digital technology, digital transactions settle instantly, reducing reliance on intermediaries such as banks and payment processors, minimising the costs associated with cash management and payment processing. This is particularly valuable for cross-border transactions, which often involve high fees and slow settlement times (Genc and Takagi, 2024; Dathe, 2025).

CBDCs provide central banks with a new tool for implementing and fine-tuning monetary policy. For example, in times of economic downturn, governments could use CBDCs to distribute stimulus payments directly to citizens, ensuring rapid and targeted financial support. With the increasing digitalisation of payments, the use of physical cash is steadily declining. CBDCs offer a government-backed digital alternative that ensures central banks remain relevant in the evolving financial landscape (Genc and Takagi, 2024; Gruin and Knaack, 2020).

While CBDCs provide new instruments for monetary policy, they also pose additional challenges. For example, if CBDCs become widely popular, central banks may need to reconsider traditional policy mechanisms, such as setting interest rates and managing the money supply, to maintain economic stability. Implementing CBDCs also requires significant regulatory adjustments, including anti-money laundering (AML) and counter-terrorism financing (CTF) measures (Atlantic Council, 2024).

Furthermore, the successful rollout of a CBDC requires robust digital infrastructure, including secure networks, reliable internet access, and integration with existing financial systems. As a fully digital system, CBDCs could be vulnerable to hacking, cyberattacks, and technical failures. Thus, ensuring the implementation of strong cybersecurity measures is crucial to maintaining trust and preventing financial disruptions. As a result, developing countries, in particular, may face major challenges in implementing CBDCs due to technological limitations (Dathe, 2025).

Government-controlled digital currencies also raise public concerns about surveillance and data privacy. Unlike cash, which provides anonymity, CBDCs could allow governments to track transactions, potentially leading to concerns about financial privacy and the misuse of personal data. As a fully digital system, CBDCs could be vulnerable to hacking, cyberattacks, and technical failures. Again, ensuring strong cybersecurity measures is crucial (Genc and Takagi, 2024).

Given these complexities, central banks worldwide are actively exploring the feasibility of CBDCs, conducting pilot programmes, research initiatives, and public consultations to evaluate their potential impact. Their goal is to balance the benefits of digital transformation – such as financial inclusion, payment efficiency, and innovation – while maintaining financial stability and consumer protection. The success of CBDCs will ultimately depend on the ability of central banks to design policies that uphold trust, security, and economic stability while embracing the opportunities of the digital economy (IMF, 2024; Dathe, 2025).

7.4 The Future of Blockchain Technology

Since the birth of Bitcoin, blockchain technology has evolved far beyond its initial use in cryptocurrencies. One of blockchain's most significant strengths is its ability to function as a decentralised, immutable distributed ledger, ensuring transparency, security, and efficiency in digital transactions. Today, blockchain technology is commonly recognised as a transformative force across multiple industries, including finance, supply chain management, healthcare, real estate, and governance. As the technology matures, advancements in scalability, regulation, and security will define its future and its role in reshaping decentralised digital economies (Chowdhury *et al.*, 2018).

In the field of finance, blockchain is revolutionising financial systems by facilitating faster cross-border payments, reducing transaction costs, and eliminating intermediaries (Catalini and Gans, 2019). DeFi has emerged as a major innovation, enabling users to access financial services such as lending, borrowing, and trading without traditional banks. However, DeFi also poses risks such as smart contract vulnerabilities, regulatory uncertainty, and cybersecurity threats (Ethereum, 2025).

Blockchain allows real-time tracking of goods and prevents fraudulent activities. Companies such as IBM and Maersk have implemented blockchain-based supply chain solutions to enhance transparency, traceability, and efficiency in global trade (IBM, n.d.; Harvard Business School, 2017).

In the healthcare sector, blockchain is increasingly used to secure patient records, streamline data sharing, and reduce fraud. By ensuring data integrity and interoperability between healthcare providers, blockchain can improve patient outcomes (Mackey, 2020).

Furthermore, the use of smart contracts in real estate transactions reduces reliance on intermediaries, lowering costs and improving transaction speeds. Blockchain also enhances property rights verification and prevents title fraud (Gutierrez and Xu, 2023).

Governments worldwide are exploring blockchain for secure voting systems, digital identities, and public records management, ensuring transparency and reducing corruption (Olalekan, 2024).

Despite its potential, blockchain technology faces several critical challenges that must be addressed for mainstream adoption. First of all, regulators need to find ways to regulate blockchain without stifling innovation. Issues include taxation, compliance with AML and CTF regulations, and the legal recognition of smart contracts (US Council on Foreign Relations, 2024).

While blockchain itself is secure, vulnerabilities exist in smart contracts and exchange platforms, leading to hacks, fraud, and financial losses. The 2021 Poly Network hack, in which attackers stole over US$600 million in cryptocurrency, highlights these risks (Ethereum, 2025). In addition, since blockchain transactions are permanent and transparent, ensuring data privacy remains a challenge, particularly for sensitive personal or business information (Chowdhury *et al.*, 2018).

In addition, current blockchain networks, especially Bitcoin and Ethereum, suffer from high energy consumption and limited transaction speeds. Innovations such as Layer 2 solutions, sharding, and PoS mechanisms aim to improve scalability. Researchers are also focusing on reducing blockchain's environmental footprint, such as adopting energy-efficient consensus mechanisms and promoting carbon offset programmes (Harichandan *et al.*, 2024).

Despite the major challenges, as research and technological development progress, there is no doubt that blockchain's potential extends beyond current applications, promising innovative solutions across industries.

With increasing efforts, governments are developing state-backed CBDCs to modernise financial infrastructure while maintaining monetary policy control. Countries such as China (Digital Yuan) and the European Union (Digital Euro) are leading CBDC initiatives (Genc and Takagi, 2024).

With the rapid advancement of artificial intelligence (AI), the combination of blockchain, AI, and the Internet of Things (IoT) can enable secure machine-to-machine transactions, autonomous supply chains, and predictive analytics (Catalini and Gans, 2019).

Blockchain technology is at the forefront of digital transformation, reshaping industries, financial systems, and governance models. While challenges related to scalability, security, and regulation persist, ongoing research and technological advancements promise a more efficient, transparent, and decentralised future. As adoption accelerates, blockchain's integration with AI, IoT, and sustainable solutions will further unlock its potential, paving the way for an innovative and inclusive digital economy.

Bibliography

Atlantic Council (2024) 'Central Bank Digital Current Tracker', March. Available at: https://www.atlanticcouncil.org/cbdctracker/ (Accessed: 13 February 2025).

Binance (2019) 'DeFi #1 – Decentralized Cryptoasset Lending & Borrowing', *Binance Research*, 6 June. Available at: https://www.binance.com/en/research/analysis/decentralized-finance-lending-borrowing (Accessed: 13 February 2025).

Blockchain Council (2024) 'Permissioned and Permissionless Blockchains: A Comprehensive Guide', 10 May. Available at: https://www.blockchain-council.org/blockchain/permissioned-and-permissionless-blockchains-a-comprehensive-guide/ (Accessed: 13 February 2025).

Catalini, C. and Gans, J.S. (2019) 'Some Simple Economics of the Blockchain', *Rotman School of Management Working Paper No. 2874598; MIT Sloan Research Paper No. 5191-16*. Available at: https://ssrn.com/abstract=2874598 (Accessed: 13 February 2025).

Chainlink (2024) 'The DeFi Ecosystem', 14 August. Available at: https://chain.link/education-hub/defi-ecosystem (Accessed: 13 February 2025).

Chowdhury, M.J. *et al.* (2018) 'Blockchain as a Notarization Service for Data Sharing with Personal Data Store', *2018 17th IEEE International Conference On Trust, Security And Privacy In Computing And Communications/ 12th IEEE International Conference On Big Data Science And Engineering (TrustCom/BigDataSE)*. New York: IEEE. doi: 10.1109/TrustCom/BigDataSE.2018.00183.

CNBC (2021) 'China has Given Away Millions in its Digital Yuan Trials. This is How It Works', 4 March. Available at: https://www.cnbc.com/2021/03/05/chinas-digital-yuan-what-is-it-and-how-does-it-work.html (Accessed: 13 February 2025).

Coinbase (2025) 'What are Stablecoins?' Available at: https://www.coinbase.com/de/public-policy/advocacy/documents/what-are-stablecoins (Accessed: 13 February 2025).

CoinGecko (2024) 'Decentralized Insurance – A Deep Dive', 22 April. Available at: https://www.coingecko.com/research/publications/decentralized-insurance-deep-dive (Accessed: 13 February 2025).

CoinMarketCap (2025) 'Top 100 Crypto Coins by Market Capitalization', 12 February. Available at: https://coinmarketcap.com/coins/ (Accessed: 13 February 2025).

Cointelegraph (2024) 'Permissioned Blockchain vs. Permissionless Blockchain: Key Differences', 19 March. Available at: https://cointelegraph.com/learn/articles/permissioned-blockchain-vs-permissionless-blockchain-key-differences (Accessed: 13 February 2025).

Dathe, T. (2025) 'Navigating the Digital Frontier: Central Bank Digital Currencies on the Example of China's E-Yuan' in Endress, T. (ed.) *Digital Project Practice for Green Finance and ESG Innovations*. Boca Raton: CRC Press / Taylor & Francis Group.

Deloitte (n.d.) 'Was sind die Chancen und Risiken der Blockchain?' Available at: https://www2.deloitte.com/de/de/pages/innovation/contents/Blockchain-Game-Changer.html (Accessed: 13 February 2025).

Ethereum (2025) 'Decentralized Finance (DeFi): What's DeFi?', 29 January. Available at: https://ethereum.org/en/defi/ (Accessed: 13 February 2025).

European Commission (2019) 'The story of Satoshi Nakamoto and the 1 million bitcoins', 21 November. Available at: https://ec.europa.eu/newsroom/cef/items/663083/en (Accessed: 13 February 2025).

Genc, H. and Takagi, S. (2024) 'A Literature Review on the Design and Implementation of Central Bank Digital Currencies', *International Journal of Economic Policy Studies*, 18, pp. 197–225. doi: 10.1007/s42495-023-00125-9.

Gruin, J. and Knaack, P. (2020) 'Not Just Another Shadow Bank: Chinese Authoritarian Capitalism and the 'Developmental' Promise of Digital Financial Innovation', *New Political Economy*, 25(3), pp. 370–387. doi: 10.1080/13563467.2018.1562437.

Gutierrez, O.C. and Xu, G. (2023) 'Blockchain and Smart Contracts to Secure Property Transactions in Smart Cities', *Applied Sciences*, 13(1), 66. doi: 10.3390/app13010066.

Harichandan, S., Kar, S. and Kumar, A. (2024) 'A Comprehensive Review of Bitcoin's Energy Consumption and Its Environmental Implications' in Maglaras, L. *et al.* (eds.) *Machine Learning Approaches in Financial Analytics. Intelligent Systems Reference Library*, vol. 254. Cham: Springer. doi: 10.1007/978-3-031-61037-0_12.

Harvard Business School (2017) 'Maersk – Reinventing the Shipping Industry Using IoT and Blockchain'. Available at: https://d3.harvard.edu/maersk-reinventing-shipping-industry-using-iot-blockchain/ (Accessed: 13 February 2025).

Hedera (n.d.) 'What are Distributed Ledger Technologies (DLTs)?' Available at: https://hedera.com/learning/distributed-ledger-technologies/what-are-distributed-ledger-technologies-dlts (Accessed: 13 February 2025).

IBM (n.d.) 'Blockchain for Supply Chain Solutions'. Available at: https://www.ibm.com/blockchain-supply-chain (Accessed: 13 February 2025).

IMF (2024) 'Central Bank Digital Currency: Progress And Further Considerations', 8 November. Available at: https://www.imf.org/en/Publications/Policy-Papers/Issues/2024/11/08/Central-Bank-Digital-Currency-Progress-And-Further-Considerations-557194 (Accessed: 13 February 2025).

Mackey, T.K. and Nayyar, G. (2020) 'Combating Health Care Fraud and Abuse: Conceptualization and Prototyping Study of a Blockchain Antifraud Framework', *Journal of Medical Internet Research*, 2(9), e18623. doi: 10.2196/18623.

Nakamoto, S. (2008) 'Bitcoin: A Peer-to-Peer Electronic Cash System'. Available at: https://bitcoin.org/bitcoin.pdf (Accessed: 13 February 2025).

OKX (2025) 'OKX State of DEX 2025 report now live: Solana retail at 48% of DEX volumes, Ethereum preferred for trades of USD50k+', 11 February. Available at: https://www.okx.com/learn/okx-state-of-dex-2025 (Accessed: 13 February 2025).

Olalekan, O.A. (2024) 'Blockchain Technology and Anti-Corruption Measures in the Setting of Public Administration in Nigeria', *African Journal of Law, Political Research and Administration*, 7(3), pp. 69–79.

Sunyaev, A. (2024) 'Distributed Ledger Technology' in *Internet Computing*. Cham: Springer.

US Council on Foreign Relations (2024) 'The Crypto Question: Bitcoin, Digital Dollars, and the Future of Money', 17 January. Available at: https://www.cfr.org/backgrounder/crypto-question-bitcoin-digital-dollars-and-future-money (Accessed: 13 February 2025).

World Economic Forum (2021) 'Cryptocurrencies Can Enable Financial Inclusion. Will You Participate?', 9 June. Available at: https://www.weforum.org/stories/2021/06/cryptocurrencies-financial-inclusion-help-shape-it/ (Accessed: 13 February 2025).

Chapter 8

Robotics and Artificial Intelligence

8.1 Robotics and Robots

8.1.1 Scope and definition

Robotics is the interdisciplinary study and practice of the design, construction, operation, and use of robots. Within mechanical engineering, robotics is the design and construction of the physical structures of robots, while in computer science, robotics focuses on robotic automation algorithms. Robotics is an important and promising technology and offers opportunities for productivity gains and thus economic growth. The core area of robotics is the development and control of robots, with recent developments taken into account in the revision and expansion for this new edition. Robots interact with the physical world based on sensors, actuators, and information processing.

A robot is an intelligent device or machine capable of sensing and interacting with its environment, utilising integrated programming options, sensors, or computer vision. Currently, AI has become a more familiar presence in robotic solutions, introducing flexibility and learning capabilities. A robot with intelligence provides new opportunities for industries to produce work safer, save valuable time, and increase productivity (Maier, 2023). Economic impact assessments and awareness of the social, legal, and ethical problems of robotics and AI are essential to optimise the advantages of these innovations while minimising adverse effects. The impact of AI and robots affects healthcare, manufacturing, transport, and jobs in logistics, security, retail, agri-food, and construction. This chapter outlines the vision of AI and the timeline of robotics development, highlighting the limitations of robots and hence embedding AI into real-world robotic applications to achieve optimised solutions (Darapureddy, 2021).

8.1.2 Operation areas

Robots are used in many fields across business, social, healthcare, and military sectors. As technology continues to develop rapidly, robots are becoming increasingly prevalent in our daily lives. They can be found in factories and other industrial settings, performing tasks that humans find dangerous or difficult to do. But robots are also starting to appear in more commonplace environments, such as homes and hospitals. While there is some concern about the increasing use of robots, it's clear that they offer a number of advantages that can benefit everyone. In this chapter, we take a closer look at some typical examples of robots in everyday life, the benefits of robots, and how they are changing human life. The goal of most robotics is to design machines that can help and assist humans (De Florio-Hansen, 2020). Many robots are built to do jobs that are hazardous to people, such as finding survivors in unstable ruins and exploring space, mines, and shipwrecks. Others replace people in jobs that are boring, repetitive, or unpleasant, such as cleaning, monitoring, transporting, and assembling. Today, robotics is a rapidly growing field, as technological advances continue; researching, designing, and building new robots serve various practical purposes (Helmold, 2024).

8.2 Evolution and Intelligence of Robots

Industrial robots are not usually humanoid in shape, although they are capable of reproducing human movements and behaviours but with the strength, precision, and speed of a machine. The first industrial robots were developed by George Devol, an American inventor and founder of Unimation, the first robotics company in history. In 1954, what is considered the first industrial robot was developed in the U.S. The first robot was a hydraulic arm called Unimate, used to lift heavy loads, which was sold to General Motors. In the following years, Ghe company developed several robot versions of the same model of the company

Unimation that were introduced gradually in some factories, mainly in the automotive sector.

It was at the end of the 1960s and in the 1970s when considerably more advanced robotic arms appeared, with cameras or sensors already incorporated. The Shakey robot, designed in 1966 by the Stanford Research Institute, stands out as an important milestone for mobile robotics. Shakey was the world's first mobile robot, thanks to software and hardware that allowed it to perceive and understand the environment, albeit in a limited way. The first mobile industrial robots also appeared in parallel. In 1954, Barrett Electronics Corporation developed the first electric vehicle that did not require a human driver, what is now known as the first autonomous guided vehicle (AGV). AGVs acquired more complex behaviours in the 1980s as technology advanced, and by the 1990s, AGVs with much more precise sensors and lasers were designed and produced. Although the first industrial robots were created in the U.S. in the 1980s and 1990s, they were already being developed in some European and Asian countries, mainly in Japan and Sweden. You probably remember IBM's Deep Blue beating world champion Garry Kasparov in a game of chess. Nowadays, the development of AI or other technologies such as those mentioned at the beginning of the chapter seems so advanced that the story about the game against Kasparov sounds outdated. The applications that AI brings to robotics, and thus to the industrial sector, are infinitely more valuable and profitable than winning chess games. Some of the advantages that AI brings to industrial robotics are as follows:

- increased accuracy in operations, processes, and quality;
- improved decision-making (especially in the face of obstacles);
- predictive maintenance.

Automation is a field in constant change, so sometimes it is difficult for some companies to embark on the path to a smart factory. For this reason, mobile robotics should always be customised and adapted to the specific needs of the company, whatever its size. Automation and robotics go hand in hand.

Figure 8.1 illustrates the evolution and maturity levels for robots. In manual processes, humans plan, execute, and control the entire process. Robots are not involved. At a higher level of maturity, i.e., the imitation level, humans plan and control work and results. The robots learn, execute

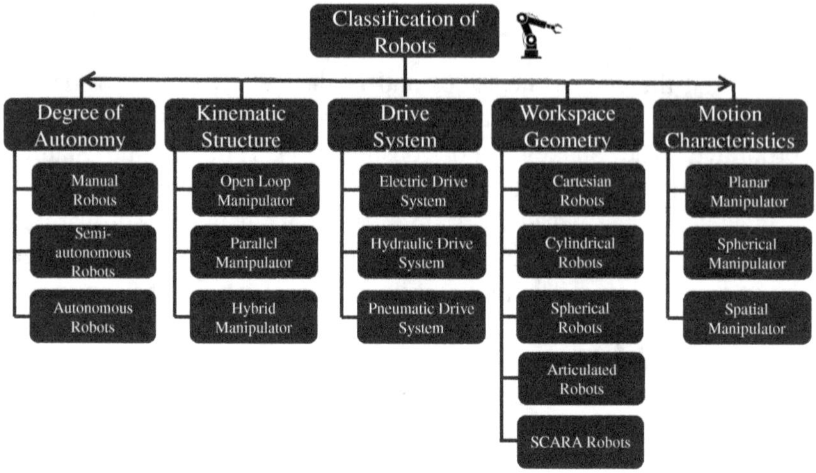

Figure 8.1. Classification of robots.

work plans and deliver. The highest maturity level is the full autonomy level. This evolution shows that robots at the highest maturity level individually learn, plan, and execute the processes. Humans only check the results on a random basis.

8.3 Classification of Robots

8.3.1 Categorisation and clustering

A robot is an intelligent artificial machine capable of sensing and interacting with its environment, utilising integrated sensors or computer vision. AI refers to science-generating devices with functions such as reasoning, thinking, learning, and planning. In the present day, AI has become a more familiar presence in robotic solutions, introducing flexibility and learning capabilities. A robot with AI provides new opportunities for industries to produce work more safely, save valuable time, and increase productivity. Economic impact assessments and awareness of the social, legal, and ethical problems of robotics and AI are essential to optimise the advantages of these innovations while minimising adverse effects. The impact of AI and robots affects healthcare, manufacturing, transport, and jobs in logistics, security, retail, agri-food, and construction. Robots can be categorised based on their degree of autonomy, drive technology, kinematical structure, degrees of freedom, workshop geometry, and movement characteristics.

8.3.2 Degree of autonomy

Figure 8.2 illustrates the classification of robots. The first category is the degree of autonomy a robot has (Robotplatform, 2023). Manual or controlled robots are robots that require human intervention to accomplish a task. They can either be wire-controlled or remotely controlled systems guided to perform any kind of complicated activities (Helmold, 2024). A remotely controlled robot can be programmed and guided to perform dangerous and complex tasks without being on the spot. Semi-autonomous robots are robots that take the best of both worlds. The intelligence built in helps them perform simple tasks and make simple decisions. For complex tasks, however, human intervention may be required. In this case, the program is designed to take intelligent decisions on its own until any human input is provided. Autonomous robots, as the name suggests, work autonomously. They can be programmed to be self-learning or humanoid. They have preprogrammed directions and are given a choice of taking decisions based on situations and surroundings. A self-learning and intelligent robot can even learn certain behaviours and act accordingly (with AI in itself being a very vast topic and the reason we will not delve into it in depth) (Kuka, 2024). Autonomous robots can be as simple as an

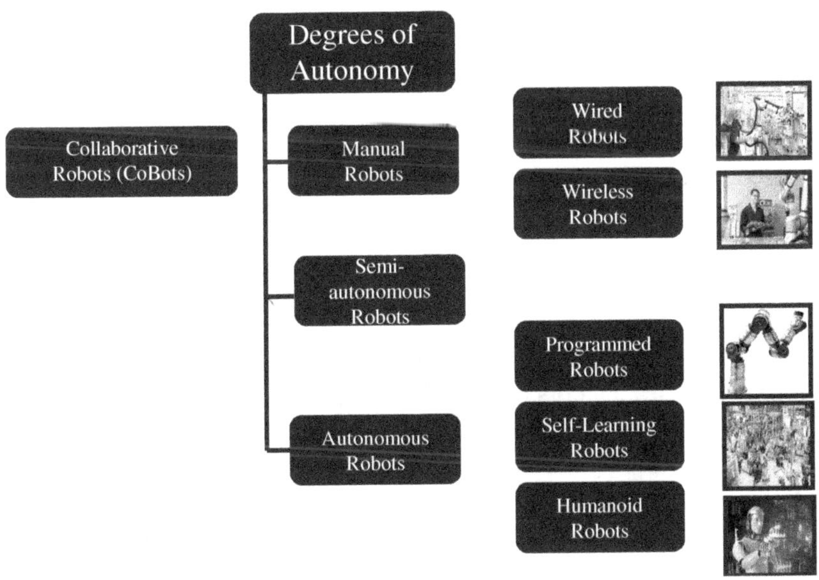

Figure 8.2. Examples of manual, semi-autonomous, and autonomous robots.

obstacle avoider or as complex as an intellectual humanoid. However, due to restrictions in power, size, and intelligence, these robots may not be good enough to perform complicated tasks (Robotplatform, 2023).

8.3.3 Kinematic structure

Kinematics pertains to the motion of bodies in a robotic mechanism without regard to the forces and torques that cause the motion. A robot's kinematic structure is described by a set of links, which for most purposes are considered to be rigid bodies, and joints connecting them and constraining their relative movement, for example, rotational or translational joints. Since robotic mechanisms are by their very essence designed for motion, kinematics is the most fundamental aspect of robot design, analysis, control, and simulation. The robotics community has focused on efficiently applying different representations of position and orientation and their derivatives with respect to time to solve foundational kinematics problems. Kinematic structures are related to manipulator systems. A robot manipulator is an electronically controlled mechanism consisting of multiple segments and performs tasks by interacting with its environment. They are also commonly referred to as robotic arms. Robot manipulators are extensively used in the industrial manufacturing sector and also have numerous specialised applications. The study of robot manipulators involves dealing with the positions and orientations of the several segments that make up the manipulators. A manipulator is called an open-loop manipulator if its links form an open-loop chain. A manipulator is called a parallel manipulator if it is made up of a closed-loop chain. A manipulator is called a hybrid manipulator if it consists of open-loop and closed-loop chains.

8.3.4 Drive system

Drive systems refer to the mechanisms, systems, or actuators that drive a robot arm. These actuators may be electric motors of some type or hydraulic or pneumatic cylinders. Hydraulic and pneumatic actuators are generally suited to driving prismatic joints since they produce linear motion directly. They are often referred to as linear actuators. Electric motors which produce rotation are more suited to driving revolute joints.

8.3.5 Workspace geometry

The volume of space where the end of the effector can reach is called the manipulator's workspace.

Cartesian Robot: A Cartesian robot has three linear joints (or a combination of them) that use the Cartesian coordinate system (*X*, *Y*, and *Z*). This robot may also have an attached wrist to allow for rotational movement. The prismatic joints deliver linear motions along the respective axes.

Spherical Robot: A spherical robot is a robot with two rotary joints and one prismatic joint. These types of robots have two rotary axes and one linear axis. Spherical robots have an arm which forms a spherical coordinate system.

Cylindrical Robots: Cylindrical robots have a rotary joint at the base and a prismatic joint to connect the links. The robots have a cylindrical-shaped work envelope, which is achieved with a rotating shaft and an extendable arm that moves in a vertical and sliding motion. Cylindrical robots are often used in tight workspaces for simple assembly, machine tending, or coating applications due to their compact design.

Articulated Robots: With articulated robots, the mechanical movement and configuration closely resemble those of a human arm. The arm is mounted to a base with a pivoting joint. The arm itself can feature anywhere from two rotary joints up to ten rotary joints, which act as axes; each additional joint or axis allows for a greater degree of motion. Most articulated robots utilise four or six axes. Typical applications for articulated robots are assembly, arc welding, material handling, machine tending, and packaging.

SCARA Robot: The selective compliance articulated robot arm (SCARA) is a type of industrial robot. By virtue of the SCARA's parallel-axis joint layout, the arm is slightly compliant in the *X*–*Y* direction but rigid in the *Z* direction, hence the term "selective compliance". This is advantageous for many types of assembly operations, for example, inserting a round pin into a round hole without binding. The second attribute of the SCARA is the jointed two-link arm layout similar to human arms, hence the often-used term "articulated". This feature allows the arm to extend into confined areas and then retract or "fold up" out of the way. This is advantageous for transferring parts between cells or for loading/unloading process stations that are enclosed. SCARAs are generally faster than comparable Cartesian robot systems. Their single pedestal mount requires a small footprint and provides an easy, unhindered form of mounting. On the other hand, SCARAs can be more expensive than comparable Cartesian systems, and the controlling software requires inverse kinematics for linear interpolated movements. However, this software typically comes with the SCARA and is usually transparent to the end user (Fanuc, 2024).

8.3.6 Motion characteristics

Planar Manipulator: The planar manipulator is named after its planar mechanism. Planar robots can be used to transport, handle, process, and test goods in a variety of industrial applications. They are much more versatile and adaptable than a conveyor belt. Planar robots are ideal for flexible production in sectors such as semiconductor manufacturing.

Spherical Manipulator: A manipulator is called a spherical manipulator if all the links perform spherical motions about a common stationary point.

Spatial Manipulator: A manipulator is called a spatial manipulator if at least one link of the mechanism possesses a general spatial motion.

8.4 Robotic Systems Architectures and Programming

Robot system architectures, robot software systems, and robot programming are complex and aim to drive robots in their environment efficiently and safely. This complexity is due, in large part, to the need to control diverse sensors and actuators in real time, in the face of significant uncertainty and noise (Siciliano, 2016). Robot systems must work to achieve tasks while monitoring for and reacting to unexpected situations. Doing all this concurrently and asynchronously adds immensely to system complexity. The use of a well-conceived architecture, together with programming tools that support it, can often help to manage that complexity. Currently, there is no single architecture that is best for all applications – different architectures have different advantages and disadvantages. It is important to understand those strengths and weaknesses when choosing an architectural approach for a given application.

8.5 Intelligent and Collaborative Robots

At the beginning of the twenty-first century, the industrial robot was reinvented as the industrial intelligent robot, able to perform highly complicated tasks like skilled production workers, mainly due to the rapid advancement in vision and force sensors. In addition, collaborative robots were developed to meet the demand for collaboration between human operators and industrial robots. The industrial intelligent robots and collaborative robots have recently been a key technology to solve issues faced by today's manufacturing industry, including a decreasing number of skilled workers and demands for reduced manufacturing costs and delivery times (Yamaguchi and Inaba, 2023).

8.6 Advantages in Manufacturing and Other Areas

Benefits of Robots and AI in Manufacturing

- **Productivity Optimisation:** It is possible to program robotics equipment to run at a constant and optimum speed without delay. This means that automated machines generate more in a shorter time than human employees. Fast settings or issues can be resolved with remote administration. Automated devices are often highly scalable and can flexibly move between tasks, increasing the productivity of the output.
- **Enhanced Quality:** Intelligent machines delete human error and have almost 100 percent accuracy, improving production quality. This, in turn, increases consumer loyalty, as faulty goods are not accessible to end-users.
- **Costs Reduced:** After a robot is able to replace hundreds of workers, factory owners can save on wages for employees. Within two years, a rapid ROI will compensate for the initial costs. Higher efficiency and speed are possible. The robots can operate in low-light conditions and do not require customer-saving climate control.

- **Safety:** Many manufacturing jobs entail elevated physical hazards and work in hazardous environments. Intelligent robots are replacing humans in dangerous conditions, mitigating accidents at the workplace and reducing the harm to workers' health. High-risk industries such as mining and fertiliser manufacturing deploy robots to prevent adverse accidents and guarantee the protection of their employees.

8.7 Laws of Robotics

Laws of robotics can refer to any set of laws, rules, or principles intended as a fundamental framework to underpin the behaviour of robots designed to have a degree of autonomy. For example:

- A robot may not injure a human being or, through inaction, allow a human being to come to harm.
- A robot must obey orders given by human beings except where such orders would conflict with the previous rule.
- A robot must protect its own existence as long as such protection does not conflict with the first or second rule.

In 2011, the Engineering and Physical Sciences Research Council (EPSRC) and the Arts and Humanities Research Council (AHRC) of the United Kingdom jointly published a set of five ethical "principles for designers, builders and users of robots" in the real world, along with seven "high-level messages" intended to be conveyed (ERPC, 2023):

- Robots should not be designed solely or primarily to kill or harm humans.
- Humans, not robots, are responsible agents. Robots are tools designed to achieve human goals.
- Robots should be designed in ways that assure their safety and security.
- Robots are artefacts; they should not be designed to exploit vulnerable users by evoking an emotional response or dependency. It should always be possible to tell a robot from a human.
- It should always be possible to find out who is legally responsible for a robot.

Bibliography

Asimov, I. (1950) *I, Robot*. New York: Gnome Press.

Darapureddy, N.D. (2021) 'A Comprehensive Study on Artificial Intelligence and Robotics for Machine Intelligence' in *Methodologies and Applications of Computational Statistics for Machine Intelligence*. Hershey: IGI Global.

De Florio-Hansen, I. (2020) *Digitalisierung, Künstliche Intelligenz und Robotik: Eine Einführung für Schule und Unterricht*. Stuttgart: UTB.

EPSRC (2024) 'Engineering and Physical Sciences Research Council (EPSRC)'. Available at: https://www.ukri.org/councils/epsrc/ (Accessed: 14 January 2024).

Fanuc (2024) 'Scara Robots'. Available at: https://www.fanuc.eu/de/en/robots/robot-filter-page/scara-series/selection-support (Accessed: 14 January 2024).

Helmold, M. (2024) *Erfolgreiche Transformation zum digitalen Champion: Wettbewerbsvorteile durch Digitalisierung und Künstliche Intelligenz*. Wiesbaden: Springer.

Kuka (2023) 'Media-Portal: Collections. Media Library'. Available at: https://assets.kuka.com/web/40b23262e3b2f790/kuka-media-pool---media-library/?viewType=grid (Accessed: 14 April 2024).

Maier, H. (2022) *Grundlagen der Robotik*. Stuttgart: VDE Verlag.

Robotplatform (2023) 'Classification of Robots'. Available at: https://www.robotplatform.com/knowledge/Classification_of_Robots/Classification_of_Robots.html (Accessed: 14 January 2024).

Rubio, F. *et al*. (2019) 'A Review of Mobile Robots: Concepts, Methods, Theoretical Framework, and Applications', *International Journal of Advanced Robotic Systems*, pp. 1–22.

Siciliano, B. (2016) *Springer Handbook of Robotics*. Heidelberg: Springer.

Yamaguchi, K. and Inaba, K. (2023) *Springer Handbook of Automation: Intelligent and Collaborative Robots*. Heidelberg: Springer.

Chapter 9

Virtual Supply Chains: The Digital Twin

9.1 Virtual "Smarter" Supply Chains

Supply and the supply chain are pivotal for the establishment, continuation, growth, and sustainability of industries. If you are unable to supply, there is no need to consider the rest of the process. The most significant challenge confronting the global population is the availability of resources, which is influenced by various factors, including conflicts, epidemics, international agreements and disputes, corporate decisions, digitalisation, and resource scarcity. In the absence of effective and lean supply chain management, there is a risk of significant disruption to production and business operations. The strategic and potentially risky nature of supply chain management can be a primary factor in the success of a business. A virtual supply chain (VSC) is a model that shows the structure and functioning of a real supply chain in a digital environment. Furthermore, this model functions as a simulation, providing high visibility of chain flows and real-time monitoring of chain steps with saved sources (mitigating waste, etc.). It is a cost-effective option for supply chain and related process risk assessment and risk mitigation. Virtual systems are an essential risk management component of digital transformation, facilitating "leanisation" from production to operation processes in various industries. By integrating all managerial and operational processes, including design, production, assembly, simulation, training, revision, maintenance and repair processes, and operation management in industries using digital twin technology and virtual systems, it is possible to achieve lean optimisation (few errors, continuous real-time monitoring, uninterrupted, lean, and low cost).

Supply chain management is the field of strategic business management with a globally integrated approach. The procurement function is a common target of almost every function in the chain and is of particular strategic importance due to its impact on overall costs and market share (Houlihan, 1985). Economic imbalances, resource constraints, intense competition, and global or regional events (such as pandemics, international disputes, or trade agreements) in the business world directly impact supply chains. Consequently, companies must reduce costs and focus on supply chain risk management. Supply chain companies must enhance their risk management capabilities, particularly in the digital domain, to ensure preparedness for any eventuality. In the current business climate, characterised by political and commercial developments, supply chain companies must demonstrate agility and resilience in the face of container scenarios, fluctuations in trade volume, and regional changes. At this juncture, the integration of digital tools, such as digital twins and AI, into managerial and operational activities has emerged as a pivotal risk management strategy. The ongoing trade tensions between the U.S. and China or general global cross-country trade relations are a key risk factor that needs to be considered when managing and tailoring global supply chains. In this regard, companies require speed, real-time monitoring, agility, resilience, and risk management tools and techniques that are digitally enabled. In the current business climate, where pressures on companies are mounting and risks to trade are increasing, the need for cost-reducing and flow-enabling digital solutions has become a top priority. These solutions are essential for maintaining uninterrupted and optimal supply chain flow, ensuring the right speed and quantity of goods, and facilitating access to markets. Supply chain management is the field of strategic business risk management with a globally integrated approach.

Digital platforms and advanced technologies provide real-time data flow and analysis in supply chain management, allowing faster and more accurate decisions to be made (Tseng *et al.*, 2021). However, coordinated decision-making processes are critical for the effective management of VSCs. Tools such as data warehouses, data mining, simulation, and optimisation techniques support decision-making in areas such as inventory, production, supply, and distribution planning (Fink *et al.*, 2005). In a virtual environment, leaders can motivate employees and improve performance by using transformational leadership capabilities in supply chain management. The characteristics of transformational leadership, such as intellectual stimulation, idealised impact, inspiring motivation, and individualised attention, can be effective in increasing employee motivation and productivity in a VSC environment (Singh and Menon, 2023).

In today's digital business environment, supply chain management is the most fundamental issue for industries and businesses. It is the basis of processes that also provide a strategic advantage. Up to the point of digitalisation, supply chain models offer distinct advantages, including greater predictability and standardisation. These models manage production, storage, distribution, and transportation processes, ensuring efficient delivery to the end user. This is achieved through the implementation of periodic planning and long-term supplier relations (Quintana and Quintana-León, 2021). The risks associated with accessing and obtaining the resources needed for the process, as well as with digital transformation, fast and error-free information flow, real-time monitoring of processes, and flexible design, have become even more important. The digital twin of the supply chain can test contingency plans by representing the state of the network (Ivanow *et al.*, 2019). In the traditional supply chain model, information flow and collaboration are often insufficient. However, in sectors where speed and flexibility are critical (aviation, pharmaceuticals, advanced technology, etc.), the "virtual supply chain" approach is seen as an important alternative (Martinez-Valencia *et al.*, 2021).

In VSCs, the management and coordination of supply chain processes are carried out by virtual representations rather than direct interaction with physical objects. Virtualisation removes fundamental constraints on space, time, and human observation (Figure 9.1). Individuals or organisations responsible for control and coordination do not need direct access or geographic proximity to the physical object they manage. It can be operated from different locations, which means that the movement of physical products from the source to the destination does not depend on the locations of the actors carrying out the control and coordination activities. Virtual objects extend beyond representing real-time situations. They can copy historical data and simulate future scenarios. In addition, these digital representations can be enhanced with sensor-generated information that cannot be perceived by the human senses. This enrichment supports more advanced capabilities in control and coordination, enabling functionality such as real-time tracking and tracing, enhanced quality monitoring, and improved planning processes (Verdouw *et al.*, 2013). A VSC is a digitalised representation of a physical supply chain that can replicate

Figure 9.1. Opportunities of virtual supply chain for optimisation.
Source: Created using Napkin AI (https://www.napkin.ai).

every aspect of the original. A VSC system is able to provide a visual representation of current operations, predict future events, and identify potential issues that may arise in the event of demand changes, shortages, and other disruptions to the supply chain. This functionality is comparable to that of a virtual reality device. The first step in implementing a VSC system is to create its digital model, also known as a supply chain digital twin. The model itself is not intelligent, but it can provide insights into the potential outcomes of changes. A digital twin, in isolation, is not a representation of the intricate processes inherent to your supply chain (Figure 9.2). It does not include your business processes, your people, and their know-how. This means it needs to have the intelligence to predict and respond in order to optimise and lean operations as the world around it continues to change (Hadavi, 2022).

Lean and agility are presented as leading general strategies in the supply chain. Lean aims to create value streams with less effort and wasted time. Agility, on the other hand, involves leveraging market knowledge and a virtual company structure to identify and seize profitable

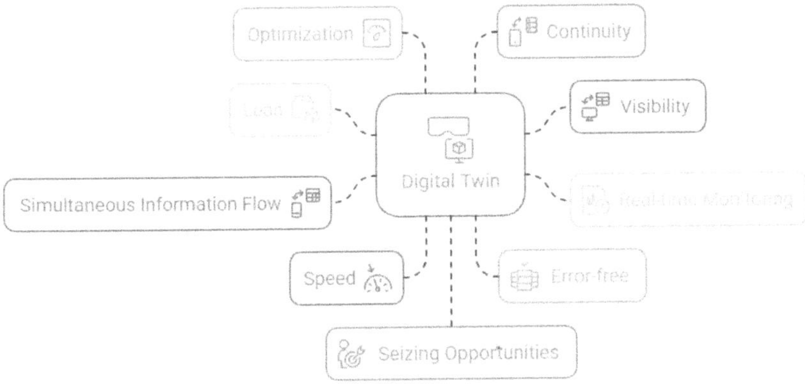

Figure 9.2. Digital twin in supply chain management.

Source: Created using Napkin AI (https://www.napkin.ai).

opportunities in a dynamic and ever-changing market (Ambe, 2010). For a supply chain to be agile, lean, and resilient, it must possess specific characteristics (Christopher, 2000):

- **Market sensitivity:** This is the ability of the supply chain to interpret and respond to actual demand, moving away from being solely forecast-driven.
- **The use of information technologies** is essential for the smooth running of the company. This refers to the process of data sharing between buyers and suppliers, which facilitates the establishment of VSCs based on information rather than inventory.
- **Process integration:** This denotes collaborative endeavours between the buyer and the seller, joint product development, shared systems, and comprehensive information exchange.
- **A network of interconnected partnerships**: It is acknowledged that individual businesses may not be able to achieve competitive advantage independently. This results in a focus on managing relationships with partners and final consumers as a network, also known as "network competition".

In the current business world, economic imbalances, resource constraints, fierce competition, and global or regional events that directly affect supply chains, such as pandemics, disputes between countries, or trade agreements, require companies to reduce costs and specialise in supply chain risk management. Supply chain companies must enhance their risk management capabilities, particularly in the digital domain, to ensure preparation for any eventuality.

A VSC can be understood as a mirrored information model of a physical supply chain. A VSC is an execution field for the simulation. It should be structured as an information model by cooperatively connecting models of component enterprises. A physical supply chain would be constructed from the cooperative and dynamic network of component enterprises in a supply chain. The implementation of a VSC as a multi-agent system involves the integration of each component enterprise as a software agent, thereby adopting an executable model.

In the field of supply chain management, software has played a pivotal role in automating operations over the past decade. Advanced planning and scheduling (APS), warehouse management systems (WMS), and transportation management systems (TMS) are prime examples of software solutions that have streamlined processes for suppliers, buyers, and carriers, facilitating seamless collaboration and efficiency. Digital twins can be integrated with existing SCM tools to provide an additional innovation layer. They facilitate enhanced data input and deliver predictive analysis for various scenarios. For instance, a global original equipment manufacturer (OEM) used a digital twin to enhance their TMS policies, achieving an 8 percent reduction in shipping and damage costs. The advantages of digital twins in the context of SCM include the following (Oca *et al.*, 2024):

- **End-to-end integration:** This integrates APS, WMS, and TMS tools to provide a comprehensive view of the supply chain. (For example, a retailer might increase efficiency in this way.)
- **Flexibility in changing markets:** This offers real-time visibility and analysis, enabling the detection of risks and the suggestion of fast solutions. (For example, an OEM has succeeded in reducing last-mile delivery costs by 5 percent.)
- **Achieving balance across multiple objectives:** This system facilitates swift decision-making between sales and operations priorities. (For example, an automotive OEM adjusted demand based on supply changes.)
- **Scenario testing:** The analysis encompasses key performance indicators such as delivery times, demand, and supplier reliability. (For example, a consumer goods company reduced warehouse costs by 15 percent.)

The VSC represents an ecosystem where digital data flow is critical in addition to physical product flow, and stakeholders constantly interact through real-time information sharing and cloud-based infrastructures. Digital, flexible, and agile competencies and capabilities highlight companies' ability to adapt to fast-changing environments because every link of the supply chain can make flexible decisions by being fed with instant data. By reducing physical stock (or with the shared stock model), costs can be reduced, and processes are accelerated (Martinez-Valencia *et al.*, 2021).

Adam Bimson, director and co-founder of Vuealta, emphasises agility and resilience as critical for businesses to stay competitive during disruptions such as COVID-19. He highlights the rise of curbside pickup and direct-to-consumer (D2C) models – as seen in companies like Nike – as adaptive strategies. Amazon's ability to leverage real-time POS data is noted as a key success factor. Rather than seeing Amazon as a threat, Bimson suggests using it as a model. He asserts that supply chains, not individual companies, now compete, and collaboration across networks is essential. He advocates for VSCs supported by real-time, event-based scenario planning to improve responsiveness and resilience (Bimson, 2021).

The adoption of digital twins is transforming supply chain efficiency by offering never-before-seen levels of visibility and predictive capabilities. Businesses can track inventory levels, production schedules, and logistics operations in real time, allowing them to quickly identify and resolve bottlenecks. Predictive analytics plays a crucial role in forecasting demand fluctuations, potential supply chain delays, and equipment failures, enabling companies to implement proactive strategies. By simulating different operational scenarios, companies can refine production schedules, optimise logistics, and reduce costs. Further, AI-powered insights can facilitate data-driven decision-making, enabling businesses to make informed choices regarding procurement, warehouse management, and transportation. Beyond efficiency, digital twins contribute to sustainability by reducing waste, optimising inventory management, and improving energy consumption (Sharma, 2025).

Cyrus Hadavi, a member of Forbes Councils, emphasises the importance of dynamic VSC systems. This demands the capacity to adapt to changes in the physical world, as well as the ability to self-correct and evolve. Just as digital twins are adapted and updated in line with physical counterparts, the digital twins of the supply chain must be able to adapt to physical changes. This concept has been termed the "metaverse of planning", where the digital model monitors changes in physical behaviour and adapts accordingly. For instance, the system detects changes in equipment efficiency or increased production costs due to spare parts utilisation and makes recommendations for improvement. In the age of e-commerce, delivery times are expected to be quick and efficient. This demonstrates that reacting to events is too slow and that predicting future consumer demands is the most effective way to fulfil these expectations (Hadavi, 2022).

Decentralised supply chain activities, complex supply chain networks, and fluctuating market environments can undermine the stability of supply chains and make them more susceptible to disruptions and risks. The sudden outbreak of the COVID-19 pandemic has made it impossible for businesses to anticipate the effects of this event. Global supply chains have faced disruptions at various levels, which has severely impacted the flow of production. In particular, manufacturing industries such as automotive, food, and pharmaceutical production, as well as service sectors such as airlines and hotels, have suffered significant losses. For example, Toyota's local supplier, Kojima Press Industries, experienced problems with its internal systems due to COVID-19 and had to shut down its servers. This supply disruption led to Toyota's ordering system becoming inoperable and all Toyota factories in Japan closing. To mitigate the impact of disruptions, companies need to reconfigure their production systems, supply chain activities, and processes. Improving supply chain resilience is recognised as an effective way to protect against disruptions and mitigate supply chain weaknesses. However, more research is needed on building supply chain resilience. The clarity of this concept and its connection with practical activities are ambiguous. It is important for businesses to focus on how specific practical activities should be linked to supply chain resilience (Zhao *et al.*, 2023).

During a panel discussion at the Morgan Stanley Laguna Conference in Dana Point, Boeing CEO Dennis Muilenburg stated that the use of digital twins had resulted in a 40 percent improvement in the quality of components and systems in aircraft manufacturing. He also emphasised Boeing's intention to expand the use of what the company terms "model-based engineering" throughout the development and simulation processes of aircraft lifecycles. It is vital to note that the primary strategy aims to digitise not only development but also the entire supply chain, including service and support operations (Freist, 2018). Improving the digital and analytical skills of the workforce is vital to supply chain transformation. The key question is how to use digital and human resources together productively. Supply chain companies compete with outside companies and different parts of their own organisation for new capabilities. This may create a gap in their ability to combine expertise and analytical skills. The future focus should therefore be on the following areas (KPMG, 2025):

- **Encourage digital/human collaboration:** Create a modern, digital work environment and let people make key decisions.
- **Make the supply chain a business partner:** Make the supply chain a business partner across functions.
- **Establish a digital centre of excellence:** Establish a digital centre of excellence and focus investment on opportunities that will have the most impact.
- **Identify new roles and supporting capabilities:** Identify new roles, capabilities, and approaches to help achieve organisational goals.

9.2 Digital Twin Concept and Integration with Industry Practices

In industries, the multiplicity and complexity of systems and processes, the difficulty of accessing and acquiring resources in every sense (tangible and intangible), and the low tolerance for error and the need for operational continuity all favour a strategy of pioneering innovative technologies that increase speed and quality, reduce error, and support optimisation. Historically, leadership styles have relied heavily on intuition, individual experience, and limited data for decision-making. The contingency approach is transforming decision-making processes, decision models, and decision support systems in the digital environment, especially with technologies such as Big Data, AI, and digital twins.

In recent years, the practice of equipping machines with sensors as part of Industry 4.0 strategies has become increasingly common. The digital twin is at the forefront of this technological revolution, thanks to its powerful data analytics and IoT integration. The IoT delivers vast quantities of data in the fields of manufacturing, healthcare, and smart cities, with significant benefits in areas such as predictive maintenance, fault detection, and traffic management. The digital twin is an innovative tool that connects the physical and virtual worlds, facilitating seamless integration between the IoT and data analytics. It provides real-time, reliable analyses to support effective decision-making processes (Parate *et al.*, 2022).

Data-driven value creation platforms are becoming central in today's economy. A key part of this is the use of digital twins – digital models that accurately reflect real physical assets and their behaviour. Digital twins help monitor the condition of a product or system by visualising real-time (or near real-time) data from physical sources such as sensors and maintenance records. This allows companies to understand the status of the asset at any given moment. To be truly useful, digital twins should do more than just store data. They need smart functions that can analyse information, build predictive models, and support future planning. For this to work, digital twins must be compatible with different systems, have user-friendly interfaces, and be designed in a neutral and independent way. A common application is the prediction of wear or failure, enabling timely maintenance and reduced downtime (Meyer *et al.*, 2020).

A digital twin creates an accurate digital model of a physical asset, ensuring continuous alignment between the digital model and the physical asset. After diagnosing, predicting, and evaluating, the twin model can send simulation results to the controller for checking the physical asset's condition, ensuring consistency between the twin and the asset. Digital twins offer advantages such as shorter design time, higher reliability, less frequent maintenance, and lower maintenance costs (Yin and Wang, 2020).

A digital twin is a dynamic and self-evolving digital model of a real-world object or entity (part, machine, process, human, etc.). This virtual model accurately represents the state of its physical twin at a given point in time, both by processing real-time data and by storing historical data. The digital twin is a precise replica of the physical twin, and any modifications made to the digital twin are reflected in the physical twin (Singh *et al.*, 2021). Digital twin is designed to enhance industry competitiveness to meet the challenges posed by variable demand, low batch size, and high-cost pressure. These factors have the potential to impact production both directly and indirectly within manufacturing. Digital twin is a technology that combines various systems to create a comprehensive digital replica of a physical asset, process or service. It integrates physical and virtual spaces (Riasco *et al.*, 2022).

The digital twin, by definition, requires a physical twin for data acquisition and context-driven interaction. The virtual system model in the digital twin can be updated in real time to reflect changes in the physical system during operation. In the contemporary business environment, the concept of the digital twin encompasses interconnected products and a digital thread, frequently facilitated by the IoT technology. The digital thread ensures seamless connectivity throughout the lifecycle of the system, facilitating the collection of data from the physical twin to ensure the continuous updating of models in the digital twin (Madni *et al.*, 2019). The key enablers of the digital twin – sensor technologies, cloud computing, Big Data, the IoT and AI – have since experienced unprecedented growth. The concept of a digital twin has recently been officially defined by NASA as a multi-physics, multi-scale, probabilistic, ultra-fidelity simulation that enables real-time replication of the state of a physical object in cyberspace based on historical and real-time sensor data (To *et al.*, 2021).

The digital twin technology plays a pivotal role in the transformation of production systems. Demonstrating that technology's potential extends beyond mere virtual representation of physical systems, digital twin enhances the flexibility, predictability, and efficiency of production processes. Digital twins use dynamic models that can be updated with real-time data to facilitate the optimisation of operations. Digital twins are used as a critical decision support tool in the design, simulation, and optimisation phases of reconfiguration processes. This integration of AI and AR technologies enables production systems to become intelligent, autonomous, and sustainable. This integrated structure has the potential to transform not only technical operation but also organisational decision-making processes, thereby making tangible contributions in areas such as energy efficiency, resource optimisation, and environmental sustainability. However, further development is required in the use of the digital twin technology, particularly with regard to data standardisation, cybersecurity, and system integration. The digital twin technology is not merely a digital tool; it is also a strategic transformation engine for manufacturing systems (Mayer *et al.*, 2025).

Although digital twins offer various advantages, it is essential to manage the associated risks. The following issues and events should be given full consideration, as they may pose potential risks. It is essential to establish and implement a comprehensive risk management system to effectively address these concerns (Fuller *et al.*, 2020):

- data security and privacy;
- infrastructure (infrastructure analysis and infrastructure capacity building) and resource management required to process and update the intensive data flow;

- standardisation, integration, and compatibility issues in inter-industry and intra-industry processes and functions;
- high costs and complex processes required to create and maintain digital twins;
- individual and organisational resistance points to cultural and organisational change and transformation;
- capacity and competence issues in relation to HR and leadership;
- problems caused by gaps and regulatory deficiencies in legal and ethical issues;
- the need for quality and accuracy of data used for the reliability of digital twins (Fuller *et al.*, 2020).

A digital twin is a digital representation of a physical asset. Real-time monitoring and control can offer several benefits, including process optimisation and the prediction of future opportunities and challenges (Newrzella *et al.*, 2021). The potential and future opportunities of digital twins are increasingly recognised by industry decision-makers and digital risk leaders. The most well understood of these benefits are listed as follows, along with the opportunities that can be seized (ORACLE, 2017):

- **Visibility:** The digital twin offers a comprehensive overview of the operational processes of machines and larger, interconnected systems, including manufacturing plants and airports.
- **Prediction:** The digital twin model can be used to predict the future state of machines by employing a range of modelling techniques, including physics-based and mathematics-based methods.

  Using carefully designed interfaces, users can easily interact with the model and pose "what-if" questions, simulating a range of conditions that are impractical to create in reality.
- **Documentation and communication mechanisms** are required to facilitate understanding and explanation of behaviour. The digital twin model can be used as a communication and documentation mechanism. This mechanism can be used to understand and explain the behaviour of a single machine or a collection of them.
- **Connecting different systems, including back-end business applications:** When designed effectively, the digital twin model can interface with back-end business applications, facilitating the achievement of business outcomes within the context of supply chain operations. These

operations include manufacturing, procurement, warehousing, transport and logistics, field service, and more.

- **To advanced analytics, detailed simulations, and informed decision-making:** Digital twins serve as virtual representations of physical systems, facilitating advanced analytics, detailed simulations, and informed decision-making through the use of the IoT (Hananto *et al.*, 2024).

The recent power outages that took place across Europe on 28 April 2025 serve to highlight the critical role of digital twins and VSCs in managing unexpected crises, in addition to the importance of digital risk leadership. Digital twins are digital representations of physical systems. They enable behaviour simulation during critical situations and enable proactive planning of alternative scenarios of the supply chain. The ability to manage supply chains based on real-time data allows companies to maintain operations even in the face of physical challenges. This is achieved by optimising and scheduling alternative logistics and production routes. It has been demonstrated that digital twins are necessary not only to enhance productivity but also to ensure operational continuity and to manage the unpredictability of challenges.

> *April 28, 2025, there was a severe power outage in the southwest of Europe. Daily life has been greatly disrupted, especially in Spain and Portugal. Regional outages were also reported in the southwest of France. This event was recorded as one of the largest power outages that Europe has experienced in recent years* (*The Guardian*, 2025).
>
> *In the Spanish cities of Madrid and Barcelona, metro services stopped, traffic lights were disabled and urban traffic experienced serious chaos* (*AP News*, 2025).
>
> *Major disruptions occurred in the communication infrastructure: mobile phone networks, internet connections and ATM systems were disabled. With the collapse of card payment infrastructure, the fragile nature of the digital economy has once again been exposed* (*Welt*, 2025).
>
> *Hospitals have tried to create temporary solutions with emergency generators, but some It has been reported that planned surgeries have been postponed and emergency departments are overwhelmed* (*AP News*, 2025).

> *It was reported that there were many people trapped in elevators and trains, and passengers had to spend the night at transport hubs in Barcelona and Madrid* (*The Guardian*, 2025).

Such large-scale outages once again highlight the effects of digitalisation and technology dependency on societies. In particular, the fact that digital payment systems, communication networks, education, modes of transport, and related infrastructure are all dependent on electricity and data grids increases the vulnerability of modern societies. It is essential to establish a risk management framework for agility.

There are three types of management approach strategies in risk management: reactive, proactive, and predictive. Each type of strategy has specific activities and behaviours that are useful in different situations. We can understand the main purpose of each type of risk management as follows (Howell, 2025):

- **Reactive:** Reduce the severity of security incidents and threats.
- **Proactive:** Identify security concerns before security incidents occur.
- **Predictive:** Predict future exposure based on past performance data.

Digital twins are a rapidly developing technology that benefits from the monitoring of complex systems, the implementation of autonomous control systems, and assistance during accidents and emergencies in real time. However, the security of DTs can be addressed by considering factors such as the safety and reliability of cyber risks. Consequently, the implementation of DTs can provide new insights for assessing risk and return (Zio and Miqueles, 2024).

Digitalisation is a proactive measure that has considerable benefits, especially in terms of predictive risk identification and management. When digital twins and VSCs are integrated with AI-based decision-making models, leaders can develop multiple strategies (e.g., risk management strategies mentioned above) to react to past events and act on present as well as forward-looking insights. AI is a valuable tool that can analyse large amounts of data in real time to predict potential risks, suggest alternative supply routes, and reprioritise operations. Digital twins are another useful technology that can help analyse risk scenarios based on data. Hybrid (agile, digital, risk-resilient) leadership plays a key role in translating these insights and strategies into actions that strengthen organisational agility and resilience. By combining AI and digital twin simulations, leaders can not only anticipate crises but also mitigate their impact before they occur. This highlights that modern supply chain leadership now relies significantly on foresight and digital intuition, just as much as on traditional planning.

The aerospace industry is experiencing significant transformation, with Airbus leading the way through the adoption of the digital twin technology. Digital twins are dynamic virtual replicas of physical objects that integrate data from design, production, and operations, enabling real-time monitoring and optimisation throughout the product lifecycle. Airbus is implementing a data-driven approach across various divisions in order to achieve end-to-end digitalisation, with a view to enhancing product development, manufacturing, and maintenance processes. Digital twins are set to transform the aerospace industry, delivering innovative, sustainable, and high-performing solutions that will shape the future of the sector. By improving design simulation, manufacturing efficiency, and predictive maintenance, digital twins will drive efficiency and sustainability in the years to come (Airbus, 2025).

As a result, digital twins are transforming decision-making across industries by providing real-time insights, predictive capabilities, and virtual simulations. This technology enables organisations to make more informed decisions, optimise processes, and improve overall efficiency in a variety of areas, including manufacturing, transportation, and smart cities (Helman, 2022; Qiu, 2024; Hananto *et al.*, 2024).

9.3 Integration of Virtual SC and Digital Twin with Industries' Processes and Systems: Best Practices

To date, digital twin applications have seen success in various industries and domains, including product design, production, prognostic and health management, and building and construction. Recent advances in sensor technologies, Big Data, cloud computing, social networks, the IoT, computer-aided design (CAD), 3D modelling, and AI have allowed the collection of massive amounts of data while enabling real-time communication for the realisation of the digital twin paradigm throughout the complete product's lifecycle (To *et al.*, 2021).

Digital twin technology is leading to revolutionary changes in various industries, with transport and smart cities being impacted. This technology creates, monitors, controls, and optimises dynamic and real-time copies of physical systems using the IoT, cloud computing, AI, and Big Data. Digital twins offer significant potential in improving battery management systems and developing smart electric vehicles, while revolutionising urban planning and solving traffic management problems. The technology also provides disaster management support by integrating a framework, improving situation assessment, decision-making, and coordination. Thanks to their ability to create 3D city models, digital twins are setting standards in transportation and urban management. Digital twins support optimised decision-making (Hananto *et al.*, 2024).

Digital twins are technologies that have the potential to transform business processes by creating a virtual representation of physical assets, with significant implications for various sectors. Digital twins are a valuable asset in a variety of business applications, including product development, design optimisation, performance enhancement, and predictive maintenance. They are of particular interest in the aerospace industry. The implementation of digital twins, particularly within specialised design, development, and critical functions, plays a major role in the development of innovative processes in the aerospace industry (Li *et al.*, 2021). Digital twin has a wide range of applications in the aerospace industry, from design and production to maintenance and operations. As the industry moves towards Industry 5.0 concepts, digital twins are increasingly integrated with other advanced technologies, such as robotics, automation, and collaborative robots (cobots). This integration will further increase their capabilities in real-time monitoring, testing, and design optimisation. This represents a significant opportunity to achieve and maintain a high level of safety and security (Prasad, 2024).

The aviation industry is one of the many industries facing a multitude of risks, which has led to a growing need for the implementation of digital twins. Digital twins can be used to manage these risks. In the context of technologically sophisticated aircraft, the digital twin concept can be employed to determine the lifetime and health of both design and materials in advance. This helps to prevent errors from causing accidents or technical issues, thereby helping to avoid material and financial losses (Wang and Yin, 2020).

Airbus is a prime example of a successful business in the aviation industry. Digital twins are at the core of their vision. A digital twin is more than a mere digital model; it is a dynamic, living virtual replica of a physical object, process, or system. This sophisticated technology integrates data from design, production, and in-service operations, providing continuous, real-time reflection of its real-world counterpart. By applying advanced analytics, simulation, and AI, digital twins enable Airbus teams to optimise processes at every stage of the product lifecycle. From the initial design and manufacturing stages through to ongoing operations and predictive maintenance, digital twin technology is transforming the aerospace industry (Airbus, 2025). In the field of aircraft design and manufacturing, digital twins are being used to facilitate virtual assembly and assembly prediction. This allows engineers to optimise complex aerospace structures. This technology also encourages technological innovation in the aerospace manufacturing sector by supporting the development of reusable spacecraft and smart rockets (Wang *et al.*, 2022). Furthermore, this technology finds implementation in the power distribution systems of commercial aircraft, enabling operational management through real-time monitoring, fault detection, and maintenance of operational efficiency (Liu *et al.*, 2023).

The process of aircraft design is multifaceted and consists of numerous steps. The digital twin concept allows for the separate acquisition of data at all stages, which can then be integrated into a digital environment. It is therefore possible to confirm that the elements required for aircraft design are brought together and that this can be achieved at a lower cost. Furthermore, the real-time data transfer facilitated by the digital twin enables immediate risk mitigation or scenario development and prediction. In this manner, the safety standards in aviation can be sustained at the required level and even enhanced. Maintaining the reliability index at the desired level is also supported by preventing unscheduled maintenance costs (Yin and Wang, 2020). It is important to note that digital risk leaders should consider incorporating these smart solutions more comprehensively into their strategies.

9.4 Smart Decision-Making: Risks and Opportunities

Decision-making involves choosing between options. However, real-world problems are often uncertain, so decision-makers must characterise their preferences in real time. Decentralised decision-making and responsiveness to changes are important factors in flexible production chains. Digital twins facilitate optimised decision-making and the development of the best strategies to improve system performance via capacity building. Digital twins, which are living digital copies of a system, offer increased situational awareness. They provide in-depth and detailed information about the current state and potential future scenarios of complex systems. This is an awareness that goes beyond traditional situational awareness. It enables "what-if" analyses, providing a framework for evaluating the potential outcomes of various decision-making options in a virtual setting (Sousa and Rocha, 2021). Digital twin technology has undergone significant development over the past 20 years, supporting digital transformation and decision-making processes in various industries. However, this rapid evolution has introduced the risk of weakening the concept's meaning and leading to ineffective applications. It is vital to improve any process or evaluate performance in organisations. Evaluating the performance of digital twins is critical to determining how effectively this technology is achieving its goals. As with any business process, it is essential to assess digital twins in order to determine their process maturity. This performance assessment is integral to ensuring the quality and reliability of the digital twin, as well as testing critical elements such as accuracy, robustness, and consistency. This will allow stakeholders to ascertain whether the digital twin accurately represents the physical system and supports decision-making processes. Consequently, these evaluation approaches assist practitioners and researchers in enhancing the effectiveness of digital twin processes. Again, digital twins facilitate optimised decision-making (Hananto *et al.*, 2024).

The challenges associated with developing a digital twin, which is a complex undertaking, vary depending on the scope and complexity of the project. The general challenges specific to digital technology are already present, as its implementation entails a number of risks, both from an engineering and commercial point of view. The risks associated with this project include a lack of understanding and appreciation of the concept in all its aspects, inadequacies regarding appropriate tools and infrastructure, high and unpredictable costs, uncertainties regarding data security, ethics, control, and transparency, as well as regulation and its enforcement and shortcomings (Singh *et al.*, 2021).

While the concept of a digital twin is applicable to many fields, there are certain factors that must be addressed to ensure its equivalent functionality to its physical counterpart. In order to achieve success with a digital twin, it is essential to establish a real-time, bidirectional, and realistic connection with the physical entity. This is made possible by accurate and efficient processing of sensor data, minimising latency, and effectively managing large volumes of data. Furthermore, as the physical asset changes, the digital twin must adapt accordingly. It is essential that decisions are transparent and easy to interpret, providing the user with an intuitive and user-friendly experience (Rasheed *et al.*, 2020).

Developing digital twin technology in aerospace is challenging due to its complexity and lack of standardisation. Better data management and virtual experimentation interfaces are needed. Research in this field is also risky to safety and increases aircraft weight. Developing the technology poses challenges. Firstly, digital twin construction is difficult, as it involves several fields and levels. The aerospace industry must establish technical architecture and digital twin construction standards. Significant progress is therefore required to move from modelling to intelligent digital twins. The absence of uniform data management specifications may cause problems in digital twin simulations and analysis. Virtual experiment implementation is also problematic, as there is no definitive interface specification for the interaction between different software programs. This makes multi-domain software difficult to develop and integrate. Conducting digital twin research in the aerospace industry carries two risks. Firstly, using AI instead of human judgement could pose a risk to aircraft operational safety. Secondly, the increased number of sensors needed for digital twins that more closely resemble physical entities could make aircraft heavier (Ying and Wang, 2020).

The aviation industry is regarded as technology-intensive due to the gradual digital transformation that has taken place over several decades within the entire aerospace industry. This has contributed to enhancing the safety and efficiency of aircraft systems. However, the integration of digital avionic systems, vehicle health management, and sensors within the aircraft has also increased complexity in terms of system configuration, maintainability, and data enormity. It is vital to emphasise the importance of effective, rapid, and accurate data management and analysis. This is in order to ensure the sustainable reliability and safety of the aircraft platform over its entire lifecycle (Li *et al.*, 2021). While digital twins offer numerous advantages, it should be noted that their implementation is not without risk and uncertainty. Recent aviation accidents, including the Boeing 737 MAX crashes, have highlighted potential issues with sensor reliability, model failure, and incorrect data processing outputs (Ibrion *et al.*, 2019).

Digital twins are a tool with the potential to improve the accuracy of monitoring, simulation, maintenance, and decision-making. By synchronising real-world behaviour with digital models, they enable more realistic interactions and enhanced data visualisation (Dani and Supangkat, 2022).

In the context of smart manufacturing systems, digital twins are instrumental in facilitating more precise and prompt decision-making processes by offering actionable insights for predictive maintenance and real-time decision-making (Feng, 2023). The application of digital twins in business decision-making extends beyond the manufacturing sector. In the context of airport operations, for instance, the implementation of digital twin technology has yielded notable benefits. It has been instrumental in optimising field resources, enhancing operational efficiency and decision-making quality by an average of 20 percent, while concurrently reducing costs by approximately 15 percent (Qiu, 2024).

Although the digital twin technology has many benefits, it is not easy to implement. Companies need to invest in the right infrastructure, including sensors, IoT devices, and data analytics tools. Partnering with a logistics provider already using this technology can help.

The market for this technology is growing. New groups are forming, such as the Digital Twin Consortium, which supports this growth and best practice. Digital twin centres are being set up in the U.K. to show the potential of digital twins in transforming maritime, aerospace, and defence.

Companies must train their staff to use and interpret digital twin data. Data security and privacy are also important, as this technology collects and processes sensitive information (Allem, 2025).

9.5 Insights and Future Issues

In today's rapidly evolving and uncertain environment, digital twins and VSCs represent more than just technological innovation – they are becoming essential tools for decision-making at the cutting edge of risk management. By enabling real-time scenario testing and predictive analytics, these systems empower leaders to make faster, smarter, and more resilient decisions. Integrating AI further enhances this capability, transforming traditional supply chains into adaptive, intelligence-driven networks. As disruptions grow more complex, the organisations that embrace these digital tools will be best positioned to navigate uncertainty and lead with confidence.

The continual evolution and rapid innovations of technology have led to significant progress in digital technology. With the development of virtual modelling and data collection technologies, the concept of digital twin has become more feasible. Digital twin platforms can improve and simulate themselves using data collected from the environment. These platforms create a digital model of the physical environment and enable real-time monitoring. Digital twin platforms also establish a connection with the physical environment for real-time tracking.

The development of virtual modelling and data collection technologies has increased the applicability of the concept of digital twin. Digital twin platforms are designed to develop and simulate themselves using data collected from the environment, thereby creating a virtual model of the physical environment and enabling real-time monitoring. Digital twin platforms also provide real-time monitoring by connecting with the physical environment. The implementation of proactive monitoring and maintenance tools has been identified as a key strategy to enhance efficiency, extend the lifespan of equipment, and reduce operational costs (Piras *et al.*, 2024).

Supply chains are using digital twins to improve long-term planning and everyday operations. Most commonly, these are used to manage inventory and goods movement, as well as production planning. However, the benefits are clear no matter how digital twins are used in both strategic and operational fields, as evidenced by the following (Oca *et al.*, 2024):

- **Strategic:** Digital twins can help with planning and other strategic decisions by simulating potential outcomes. This gives confidence to companies. For example, a retailer used a digital twin of its distribution network to test implications of a new design. Teams found they could relocate and resize a cross-dock with 50 percent less real estate.
- **Operational:** Digital twins can optimise daily decision-making across different operations. Rather than using many predictive models, a digital twin combines many models to make integrated forecasts and suggestions for operational changes. The same retailer optimised its cross-dock and daily inventory decisions. (Oca *et al.*, 2024)

As supply chains grow more complex, organisations are increasingly turning to advanced technologies to enhance efficiency, resilience, and decision-making. Among these innovations, digital twins have emerged as a transformative force in supply chain management. These virtual replicas of physical assets, processes, and systems leverage real-time data, AI, and predictive analytics to optimise operations. By providing a dynamic, data-driven approach, digital twins help businesses anticipate disruptions, streamline logistics, and improve overall performance (Sharma, 2025).

Digital twin technology is poised to transform supply chains and integrated logistics providers, offering unprecedented visibility, predictive capabilities, and efficiency. By leveraging this innovative technology and working with integrated logistics providers involved in this space, companies can optimise their operations, enhance decision-making, and build more resilient and sustainable supply chains. As technology continues to evolve, its adoption is likely to become increasingly widespread, driving significant improvements across the industry (Allem, 2025). Digital twins allow you to model and stress test potential breaking points in your supply chain and take action to improve overall resiliency. You will be able to identify and quantify risks across your network, analyse and compare results from numerous scenarios on mitigating the risk, and choose the best response for the business (Coupa, 2024).

Digital twins represent a cutting-edge technology that currently benefits from the monitoring of complex systems, the implementation of autonomous control systems, and assistance during accidents and emergencies in real time. However, aspects such as security, cybersecurity, and reliability of digital twins are still open issues that have not yet been comprehensively addressed. These aspects can provide new perspectives on the evaluation of risks and returns from the implementation of digital twins, with the identification and analysis of opportunities and threats remaining a primary focus area. From education to simulation, from processes to production, from cities to health and medicine, artificial twin and AI, which has the chance to be used in all managerial systems and activities, has become one of the essential parts of decision-making. Within the scope of the digital twin, which is an area worth investing in, it is important for decision-makers to choose the type that best suits their business and has opportunities that they can seize.

Now, our twin questions are as follows (Figure 9.3):

"Where's your digital twin?"

"Are you ready to create your digital twin?"

Figure 9.3. Twin questions.

Source: Created using Napkin AI (https://www.napkin.ai).

Bibliography

Airbus (2025) 'Innovation: Digital Twins: Accelerating Aerospace Innovation From Design to Operations', 28 April. Available at: https://www.airbus.com/cn/newsroom/stories/2025-04-digital-twins-accelerating-aerospace-innovation-from-design-to-operations (Accessed: 28 April 2025).

Allem, N. (2025) 'Digital Twins for Efficient Supply Chains'. Available at: https://www.maersk.com/insights/digitalisation/2025/04/11/digital-twins-for-efficient-supply-chains (Accessed: 11 April 2025).

Bimson, V.A. (2021) 'Virtual Supply Chains are the Key to Achieving Agility and Resiliency', *The Retail Touchpoints Network*. Available at: https://www.retailtouchpoints.com/topics/supply-chain-sourcing/virtual-supply-chains-are-the-key-to-achieving-agility-and-resiliency (Accessed: 2021).

Coupa (2024) 'The Complete Guide To Supply Chain Digital Twins'. Available at: https://www.coupa.com/blog/the-complete-guide-to-supply-chain-digital-twins/ (Accessed: 2024).

Dani, A.A. and Supangkat, S.H. (2022) 'Combination of Digital Twin and Augmented Reality: A Literature Review', *2022 International Conference on ICT for Smart Society (ICISS)*. doi: 10.1109/ICISS55894.2022.9915160.

Feng, Q. *et al.* (2023) 'Multi-Level Predictive Maintenance of Smart Manufacturing Systems Driven by Digital Twin: A Matheuristics Approach', *Journal of Manufacturing Systems*, 68, pp. 443–454. doi: 10.1016/j.jmsy.2023.05.004.

Fink, A., Sebastian, H.-J. and Voß, S. (2005) 'Intelligent Decision Support for E-Logistics and Supply Chain Management', *Proceedings of the 38th Hawaii International Conference on System Sciences*. Big Island, HI: IEEE. doi: 10.1109/HICSS.2005.333.

Freist, R. (2018) 'Boeing Aims to Virtualize the Development of Aircraft', *Hannover Messe*, 2 October. Available at: https://www.hannovermesse.de/en/news/news-articles/boeing-aims-to-virtualize-the-development-of-aircraft (Accessed: 2 October 2018).

Fuller, A. *et al.* (2020) 'Digital Twin: Enabling Technologies, Challenges and Open Research', *IEEE Access*, 8, pp. 108952–108971. doi: 10.1109/ACCESS.2020.2998358.

Hadavi, C. (2022) 'The Metaverse of Supply Chain Planning: Creating Virtual Supply Chains', *Forbes Technology Council*, 12 July. Available at: https://www.forbes.com/councils/forbestechcouncil/2022/07/12/the-metaverse-of-supply-chain-planning-creating-virtual-supply-chains/ (Accessed: 12 July 2022).

Hananto, A.L. *et al.* (2024) 'Digital Twin and 3D Digital Twin: Concepts, Applications, and Challenges in Industry 4.0 for Digital Twin', *Computers*, 13(4), 100. doi: 10.3390/computers13040100.

Helman, J. (2022) 'Digital Twin-Driven Approach Towards Manufacturing Processes Support', *15th Global Congress on Manufacturing and Management (GCMM 2021)*. Bristol: IOP Publishing.

Houlihan, J.B. (1985) 'International Supply Chain Management', *International Journal of Physical Distribution & Materials Management*, 15(1), pp. 22–38. doi: 10.1108/eb014601.

Howell, C. (2025) 'Difference Between Reactive, Proactive and Predictive Risk Management in Aviation SMS'. Available at: https://aviationsafetyblog.asms-pro.com/blog/understand-reactive-predictive-and-proactive-risk-management-in-aviation-sms (Accessed: 30 October 2025).

Ibrion, M., Paltrinieri, N. and Nejad, A.R. (2019) 'On Risk of Digital Twin Implementation in Marine Industry: Learning from Aviation Industry', *International Maritime and Port Technology and Development Conference and International Conference on Maritime Autonomous Surface Ships*. Bristol: IOP Publishing. doi: 10.1088/1742-6596/1357/1/012009.

Ivanov, D. *et al.* (2019) 'Handbook of Ripple Effects in the Supply Chain' in *International Series in Operations Research & Management Science*, vol. 276. Berlin: Springer, pp. 309–332. doi: 10.1007/978-3-030-14302-2_15.

KPMG (2025) 'The Transforming Competencies of Supply Chains', 5 May. Available at: https://kpmg.com/tr/tr/home/insights/2020/03/tedarik-zincirlerinin-donusen-yetkinlikleri.html (Accessed: 5 May 2025).

Li, L. *et al.* (2021) 'Digital Twin in Aerospace Industry: A Gentle Introduction', *IEEE Access*, 10(4), pp. 9543–9562. doi: 10.1109/ACCESS.2021.3136458.

Liu, W. *et al.* (2023) 'A Digital Twin Method for Civil Aircraft Power Distribution System Based on Unity3D and Simulink', *Journal of Physics: Conference Series*, 2615. doi: 10.1088/1742-6596/2615/1/012017.

Madni, A.M., Madni, C.C. and Lucero, S.D. (2019) 'Leveraging Digital Twin Technology in Model-Based Systems Engineering', *Systems*, 7(7), 7. doi: 10.3390/systems7010007.

Martinez-Valencia, L., Garcia-Perez, M. and Wolcott, M.P. (2021) 'Supply Chain Configuration of Sustainable Aviation Fuel: Review, Challenges, and Pathways for Including Environmental and Social Benefits', *Renewable and Sustainable Energy Reviews*, 152, 111680. doi: 10.1016/j.rser.2021.111680.

Mayer, A. *et al.* (2025) 'Digital Twins, Extended Reality, and Artificial Intelligence in Manufacturing Reconfiguration: A Systematic Literature Review', *Sustainability*, 17(5), 2318. doi: 10.3390/su17052318.

Meyer, H. *et al.* (2020) 'Development of a Digital Twin for Aviation Research', *Deutsche Gesellschaft für Luft- und Raumfahrt - Lilienthal-Oberth e.V.* doi: 10.25967/530329.

Newrzella, S.R., Franklin, D.W. and Haider, S. (2021) '5-Dimension Cross-Industry Digital Twin Applications Model and Analysis of Digital Twin Classification Terms and Models', *IEEE Access*, 9, pp. 131306–131321.

Oca, A. *et al.* (2024) 'Digital Twins: The Key to Unlocking End-to-End Supply Chain Growth'. Available at: https://www.mckinsey.com/capabilities/quantumblack/our-insights/digital-twins-the-key-to-unlocking-end-to-end-supply-chain-growth (Accessed: 31 December 2024).

Oracle (2017) *Digital Twins for IoT Applications: A Comprehensive Approach to Implementing IoT Digital Twins*. Available at: https://infotech.report/Resources/Whitepapers/a3323742-5fee-4df3-89fd-d9aeaa012ba6_digital-twins-for-iot-apps-wp-3491953.pdf (Accessed: 15 October 2024).

Parate, S., Bhivgade, A. and Bharati, M. (2022) 'Digital Twin Technology Transforming Modern Healthcare Industry', *International Journal for Research in Applied Science & Engineering Technology (IJRASET)*, 10(5), pp. 1765–1771. doi: 10.22214/ijraset.2022.42604.

Piras, G., Agostinelli, S. and Muzi, F. (2024) 'Digital Twin Framework for Built Environment: A Review of Key Enablers', *Energies*, 17(2), 436. doi: 10.3390/en17020436.

Prasad, G. (2024) 'Investigation of Aviation Sector in Industry 5.0' in *AI and Blockchain Optimization Techniques in Aerospace Engineering*. Hershey: IGI Global, pp. 135–148.

Qiu, Y. (2024) 'Digital Twin Based Resource Optimization and Dynamic Decision Making in Air Terminal Yards', *Applied Mathematics and Nonlinear Sciences*, 9(1), pp. 1–23. doi: 10.2478/amns-2024-0871.

Quintana, B.M. and Quintana-León, M. (2021) 'Differences Between a Traditional Supply Chain and an Integrated Supply Chain' in Senhoras, E.M. (ed.) *Administração: estudos organizacionais e sociedade*. Atena Editora, pp. 35–44. doi: 10.22533/at.ed.642211711.

Rasheed, A., San, O. and Kvamsdal, T. (2020) 'Digital Twin: Values, Challenges and Enablers From a Modeling Perspective', *IEEE Access*, 8, pp. 21980–22012.

Riasco, R.E.O., Sagot, J.-C. and Stjepandić, J. (2022) 'Conceptual Approach for a Digital Twin of Medical Devices' in *Transdisciplinarity and the Future of Engineering*. Amsterdam: IOS Press, pp. 320–329.

Sharma, U.K. (2025) 'Digital Twins In The Supply Chain: Transforming Operations With Real-Time Simulation And AI', *Forbes*, April. Available at: https://www.forbes.com/councils/forbestechcouncil/2025/04/29/digital-twins-in-the-supply-chain-transforming-operations-with-real-time-simulation-and-ai/ (Accessed: 29 April 2025).

Singh, G. and Menon, F. (2023) 'Impact of Transformational Leadership on Employee Motivation and Performance in Supply Chain Management', *Conference proceedings of the Centre for Business & Economic Research, ROGE-2023*, 14, pp. 140–148.

Singh, M. *et al.* (2021) 'Digital Twin: Origin to Future', *Applied System Innovation*, 4(2), 36. doi: 10.3390/asi4020036.

Sousa, M.J. and Rocha, Á. (2021) 'Decision-Making and Negotiation in Innovation & Research in Information Science', *Group Decision and Negotiation*, 30, pp. 267–275. doi: 10.1007/s10726-020-09712-z.

To, A. *et al.* (2021) 'Drone-based AI and 3D Reconstruction for Digital Twin Augmentation' in Meiselwitz, G. (ed.) *International Conference on Human-Computer Interaction.* Cham: Springer International Publishing, pp. 511–529.

Tseng, M.-L. *et al.* (2021) 'A Cause and Effect Model for Digital Sustainable Supply Chain Competitiveness under Uncertainties: Enhancing Digital Platform', *Sustainability*, 13(18), 10150. doi: 10.3390/su131810150.

Verdouw, C., Beulens, A. and Vorst, J.v. (2013) 'Virtualisation of Floricultural Supply Chains: A Review from an Internet of Things Perspective', *Computers and Electronics in Agriculture*, 99, pp. 160–175. doi: 10.1016/j.compag.2013.09.006.

Wang, L. and Yin, Z.H. (2020) 'Application and Development Prospect of Digital Twin Technology in Aerospace', *IFAC PapersOnLine*, 53–55, pp. 732–737.

Wang, X. *et al.* (2022) 'Research on the application of digital twin in aerospace manufacturing based on 3D point cloud', *2022 International Conference on Electronics and Devices, Computational Science (ICEDCS)*. doi: 10.1109/ICEDCS57360.2022.00076.

Yin, Z. and Wang, L. (2020) 'Application and Development Prospect of Digital Twin Technology in Aerospace', *IFAC-PapersOnLine*, 53(5), pp. 732–737. doi: 10.1016/j.ifacol.2021.04.165.

Zhao, N., Hong, J. and Lau, K.H. (2023) 'Impact of Supply Chain Digitalization on Supply Chain Resilience and Performance: A Multi-Mediation Model', *International Journal of Production Economics*, 259, 108817. doi: 10.1016/j.ijpe.2023.108817.

Zio, E. and Miqueles, L. (2024) 'Digital Twins in Safety Analysis, Risk Assessment and Emergency Management', *Reliability Engineering & System Safety*, 246, 110040. doi: 10.1016/j.ress.2024.110040.

Chapter 10

AI Impact on Leadership and Human Resources

A good leader always gives everyone a chance.

— M.E.YM. Efe Yilmaz, A. Kucuk Yilmaz's son

Innovation distinguishes between a leader and a follower.

— Steve Jobs, Apple co-founder

10.1 Evolutional Transformation in the AI Age: "Hybrid Leadership" and AI Integration

We strongly need new holistic and updated approaches to leadership and human resources (HR) in the digital era while managing organisational agility and resilience. Our world is undergoing constant change in terms of its economy, climate, atmosphere, and living resources. Individuals, in turn, are changing in parallel with these global shifts. A transformation is inevitable when basic dynamics change.

In the current business climate, leaders are under increasing pressure to consider the impact of digital technologies on a wide range of issues, including leadership, HR, supply chain management, organisational behaviour, ethics, control, and auditing. When considering the extensive and varied impact of these innovations on businesses, people and all organisational processes, the key resilience and agility risk strategy is to capitalise on their opportunities. In today's business world, it is imperative to understand the impact of digitalisation on productivity, organisational culture, ethical standards, and management approaches. This transformation is not only driven by leaders but also by employees and new-generation professionals. As artificial intelligence (AI) continues to evolve, its impact on the travel industry is becoming increasingly evident. From optimising travel planning to enhancing sustainability, AI is transforming the way we experience the world (Johnson, 2024).

Technological developments, especially in the fields of AI, natural language processing (NLP), digital twins, and virtual systems, are leading to unpredictable changes in every field. The most significant effects of this transformation are evident in critical areas such as business, leadership approaches, and HR management. In industries with high safety requirements, such as aviation, the integration of these technologies is reshaping operational efficiency and safety standards. Therefore, it is vital that we update our approach to leadership, human resource management, and business systems in a digitalised world, embracing a digitally driven agile and lean strategy.

The basic needs of leaders are to develop their characteristics within the scope of the big five personality model as well as their no-tech skills for empowerment within the scope of AI, to be digital/technologically literate enough to effectively incorporate digital technologies into their leadership practices, and to be digital risk managers.

AI refers to systems that can mimic human intelligence, perceive their environment and make decisions based on situations, and have the ability to learn. AI systems are capable of demonstrating the ability to solve complex problems, recognise patterns, make predictions, and recommend decisions through techniques such as machine learning, deep learning, and NLP. These features enable AI to facilitate data-driven decision-making processes with greater speed and accuracy than is possible for humans (Kumar *et al.*, 2024).

AI can actually be considered a key component in filling the gap in the human–machine interface, that is, to understand the speed and scope of developing technology and to include it in our lives. Thinking of it as a tool to keep up with the accelerating business world represents the idea that most people want to believe. It will also be possible to make life easier with a system that can model and create decision options very quickly and accurately regardless of the complexity of the problem. A care service for people in need of care (children, the elderly, the sick, etc.), a support for students for their lessons, exams, or foreign language homework, a helper for professional people for their work at work, a holiday planner, a problem solver, an alternative creator, sometimes even a psychologist, a crisis solver, or a smart and fast assistant that can be accessed from anywhere at any time – this is like a fairy tale come true. However, it is important to maintain a realistic perspective. As is widely recognised, risks have a dual potential, as both opportunities and threats. The yin–yang symbol represents the duality of all things, illustrating that every aspect of life has two sides. At the end, risks have a two-way (opportunity–threat) potential. Yin and yang – every coin has two sides. ☯

AI has the potential to provide significant benefits in supporting leadership decisions. When AI is incorporated into systems, it can support decisions ranging from strategic planning and human resource management to crisis response and organisational communication. The technical performance of these systems is not the only controversial aspect; the ethical implications, accountability, and explainability of their decisions are also a cause for concern. The use of AI in leadership is not only a matter of technology; it is also a problem of ethical governance (Kuzmanko and Vrbová, 2024).

Leadership and decision-making processes are shifting rapidly from traditional methods to digital platforms. The analytical tools offered to executives by digital twins and AI-based systems are transforming strategic decision-making processes while, at the same time, reshaping leadership styles and organisational structures. While AI-powered data analysis enables leaders to make more data-driven and objective decisions, the question of how these technologies interact with managers' intuitive decision-making abilities is still a matter of debate. In addition, the role of NLP-based feedback systems in empowering leaders to effectively manage team dynamics and conflict is becoming increasingly significant.

AI and digital twin systems have transformed the role of leaders by providing data-driven, predictive, and continuously adaptive decision support mechanisms. Leadership is no longer based on experience and intuition alone; it has become a process of strategic reasoning supported by Big Data analysis, algorithmic forecasting, and scenario-based simulations. This transformation has brought about radical changes in both decision-making processes and leadership styles.

However, human resource management is also subject to the widespread effects of digital transformation. AI-based recruitment systems are fundamentally changing the way we analyse candidates and assess their capabilities. However, a key question is how to align these technologies with ethical factors such as diversity and fairness. In the aviation industry, the use of digital twins to monitor employee behaviour and safety risks allows for higher safety standards. However, it is also critical to identify the ethical boundaries of these technologies.

AI and digital twins are transforming not only individual leadership and HR processes but also organisational structures and business processes. The integration of virtual systems offers significant opportunities, particularly in areas such as maintenance, operational simulation, and crisis management. However, this transformation also introduces new risks, including "system fatigue" and "information overload". Another important issue that requires careful analysis is the impact of digital systems on cross-cultural management and decision-making processes, especially in multinational organisations.

Our world is undergoing continuous change, both economically and in terms of climate, atmosphere, and natural resources. Individuals, in turn, are also evolving in response to these global transformations. When fundamental dynamics shift, transformation becomes inevitable. Since the era of Taylorism, the concepts of management and leadership, along with related approaches, have undergone significant evolution. In the business world, objectives have expanded and diversified, leading to the emergence of efficiency as a dominant production model aimed at achieving optimal outcomes.

It is important to note that both business and environmental conditions have changed substantially over time, including access to and acquisition of resources. Consequently, when the dynamics of the business world and human behaviour change, leadership approaches must also be revisited, updated, or reinvented.

Business strategies and models, work practices, organisational environments, communication channels, and the speed of information exchange have all accelerated, with a growing range of available options. These developments raise a key question: *Which HR strategies and leadership styles are most effective in this rapidly evolving environment?*

In the volatile, uncertain, complex, and ambiguous (VUCA) business environment, leadership is undergoing a radical transformation due to the rapidly evolving impact of AI. VUCA has gained popularity in the business world as a management buzzword. It is a way of saying, "Hey, it's crazy out there!" However, it is important to note that the term can be misleading. VUCA conflates four distinct types of challenges that demand four distinct types of responses. This can make it challenging to determine the most effective approach to a situation, potentially leading to a reliance on the VUCA framework as a crutch, rather than utilising it as a catalyst for strategic and planning efforts (Bennett and Lemoine, 2025).

The business world is increasingly reliant on agile and resilient leadership, which is why it is vital to integrate continuous and technological learning with digital culture. However, if the potential negative impacts of AI are given undue consideration, there is a risk that business objectives and decision-making processes may be affected. The dynamics of HR management are changing fundamentally. Concurrently, the field of AI is demonstrating a capacity to enhance efficiency, effectiveness and productivity, despite the presence of resource constraints and limitations in various industries. AI has become a transformative force that is influencing all areas of industry. This has resulted in a variety of opportunities for companies to develop their business and improve their competitiveness (Assis, 2024).

In the current business climate, characterised by technological and logical developments, there is an increasing need to rethink leadership, placing greater emphasis on agile and digital aspects. This new approach calls for leaders to embody the optimal integration of digital and human competencies, underpinned by a risk-oriented philosophy, within their leadership styles. In this new era, it is essential for leaders to tap into their unique core capabilities, such as strategic thinking and emotional intelligence, to leverage them as a source of strength. AI should be used as a tool to support these strengths and enable leaders to achieve more efficient results (Brown, 2024).

The combination of hybrid agile digital resilience risk-based (ADiReR) leadership with data- and risk-driven decision-making, optimisation, automation, and risk forecasting capabilities provided by AI has emerged as a style that gives organisations a sustainable and competitive advantage. AD leadership integrates motivational HR practices, direction, and organisation, which are essential for navigating the dynamic business landscape, particularly in the context of AI. This innovative leadership approach facilitates effective organisational direction, making it a valuable asset for businesses operating in today's highly dynamic and volatile business environment. AD leadership integrates a range of modern approaches, including lean leadership, digital leadership, and risk leadership. The concept of leadership has been expanded to include the qualities of inspiration and the ability to motivate through the power of following, elements that have been integrated into the framework of digital leadership. This integration enables the realisation of digital transformation and the transformation of HR processes in the present day. We have provided our professional insights on this topic, leveraging our extensive experience and expertise in management via this book.

We are happy to be able to share what we know about this. We can use our knowledge of management, as explained in this book, thanks to our leader, Prof. Dr. Marc Helmold. Also, we think and feel that human leaders are always special. Our unique dream team leader is Prof. Dr. Helmold. He is innovative, fast of fast and exactly project creator. We sure that, none and no AI (or any digital technology or system) can lead/manage/inspire/bring us together the way he can. He gives support to all team members, supports our motivation, sets roles efficiently and easily, inspiring us. Always he has brilliant project ideas to realize. We want to do our best thanks to his leadership. We are committed to achieving excellence under his guidance. We feel huge respect, and feel very lucky ourselves since we are working with him. This is definition concept of our leader (see Figure 10.1).

Digital culture has the power to transform attitudes and behaviours at both the individual and organisational levels. It introduces new work models, business techniques and tools, and changes to processes and organisational structures. Consequently, companies and their employees adopt new norms and habits. The integration of an ADiReR approach to leadership with the data-driven decision-making, automation, and forecasting capabilities enabled by AI has been shown to provide organisations with a competitive advantage. In this transformation, it is essential for leaders to adopt an approach based on flexibility, adaptability, and collaboration. In the context of competency-based HR management, the skills of future leaders will include emotional intelligence, strategic thinking, and change management, as well as technical knowledge. Successful leaders in AI-enabled digital ecosystems must possess a vision that optimises human and machine interaction, fosters innovation, and prioritises employee development. It is imperative for business

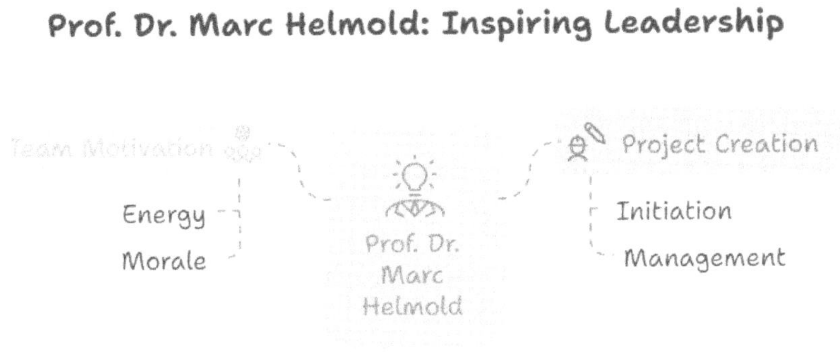

Figure 10.1. Our team and leadership.

Source: Created using Napkin AI (https://www.napkin.ai).

leaders and decision-makers to establish sustainable motivation and encourage individuals to unlock their creative potential. AD leadership motivates HR and fosters the collaboration necessary to achieve goals while also creating flexible and unrestrained spaces for individual and team creativity. Hybrid leaders transform individual uniqueness into the power of creativity for sustainable, supra-competitive advantage in businesses.

> *Intelligence is the ability to adapt to change.*
>
> — Stephen Hawking

> *Agile is all about adapting to change; it was built on the foundational principle that business drivers will change and the development teams must be ready to adapt.*
>
> — Digital.ai (2025)

A hybrid leader's capacity to inspire is contingent on their adeptness in assuming three pivotal roles: visionary, exemplar, and mentor. People around the world prefer leaders who inspire by pointing to these roles rather than leaders who exasperate them because inspirational leaders are visionaries: They take a long-term view and offer a positive, meaningful vision for the future. This fulfils people's need for meaning and purpose. Inspirational leaders serve as exemplars of desired behaviour:

They embody composure, courage, passion, competence, and humility. This fulfils the human need for protection and passion. Finally, inspirational leaders are exemplary mentors: They empower, encourage, and empathise with others, but they also challenge others to be the best version of themselves. This fulfils the human need for support and status. All of us have the capacity to develop the qualities required to be inspirational leaders in all three dimensions. Effective mentoring entails a deep understanding of individuals' diverse needs and the ability to attentively listen to these needs (Galinsky, 2025). AI integration transforms organisational cultures and leadership paradigms in today's rapidly changing digital economy. Reciprocal symmetry provides an intriguing lens for examining the changing interactions between technology, culture, and organisational leadership. AI and reciprocal symmetry impact company culture and leadership in the digital economy (Maddula, 2018).

Finally, it is clear (undoubtedly) that leadership will always be a core issue and need in any given age and circumstance, representing both a threat and an opportunity for society. The question that must now be answered is who will lead AI (or "robots", as it is otherwise known and which refers to digital technology), or will AI lead both us and decision-making.

10.2 Algorithms Rule, Who Will Be the Leader? Digitalisation, Unemployment, and Loneliness: Leadership in the Age of Robots

Leaders must be close enough to relate to others, but far enough ahead to motivate them.

— John C. Maxwell

Digitalisation has had a considerable impact on individuals, resulting in feelings of loneliness and decreased motivation. The increasing use of AI has led to concerns about job security, as people worry that automation may lead to the replacement of human workers. This issue has become a subject of debate and has been portrayed in various media, including movies and books. However, the increasing use of autonomous systems in recent years has provided real-world examples of this transformation. It is becoming evident that certain roles are becoming more precise and efficient as a result of automation. Conversely, advancements in AI and digital technology are also generating new employment opportunities. In order to successfully navigate this transition, it is essential to manage and direct public perceptions of digital transformation by fostering a robust digital culture. This is an ambitious goal that will require the leadership skills of the person tasked with its delivery.

Despite the numerous advantages offered by AI, there are significant concerns regarding its potential impact on society. As AI technologies evolve, there are several critical issues that must be given full consideration, including ethical considerations, the potential for job losses, and the risk of increased inequalities (Farahani and Ghasemi, 2024). The complexity of AI's decision-making processes poses a significant challenge to accountability and trust. This uncertainty can make it challenging for users and stakeholders to understand how the systems work, which can potentially erode trust in these technologies (Bærøe *et al.*, 2020). In addition, while AI systems are increasingly being adopted in various industries, the legal frameworks governing the use of these systems often lag behind technological developments, creating loopholes that can be exploited for malicious purposes. This poses serious challenges in terms of both legal compliance and ethical responsibility and may pose risks to society (King *et al.*, 2019).

Let us consider this issue from a different perspective. While AI has the potential to create isolated work environments, when used effectively, it can also enhance individual performance, boost creativity, and foster a sense of connection. Some professionals are now using AI as an assistant or teammate in their work. By leveraging AI as an assistant or collaborator, professionals can enhance their thinking, brainstorming, production, and inspiration, leading to a sense of empowerment. The sense of support and collaboration that this fosters can help to reengage individuals who feel tired, exhausted, or stuck and encourage a positive sense of belonging.

The rapid advancements in AI and digitalisation are transforming all aspects of our social, cultural, and economic lives. People are adapting their behaviours and attitudes to these rapid changes. Consequently, the evolving business environment is also impacting HR. This dynamic environment necessitates a reactive approach. In nature, all systems are interconnected, as open systems are inherently susceptible to external influences. In today's environment, the way in which human beings, their lives, behaviours, business practices, and communication methods, as well as how they are influenced, are all affected by AI and digitalisation is of particular interest. As an open system, human beings are influenced by the environment and its dynamics. This phenomenon has been consistent throughout history, and humans have adapted to it continuously. AI has the potential to reshape industries, transform labour markets, and influence geopolitical dynamics, raising profound questions about the future of humanity. However, by adopting a proactive and responsible approach to AI management, we can mitigate risks, maximise benefits, and ensure that AI remains aligned with human values and interests. In order to effectively manage AI-related threats, the following strategies should be implemented (Farahani and Ghasemi, 2024):

- the setting of clear ethical guidelines,
- interdisciplinary collaboration,
- transparency and accountability in AI development processes,
- regulatory oversight,
- public engagement,
- continuous monitoring and evaluation of AI systems.

Implementing these strategies will allow us to protect against the potential risks of AI while also capitalising on its transformative potential.

The integration of AI and digital technologies creates a delicate dilemma: While it can ease and enhance certain aspects of life, it simultaneously fosters individualism and isolation. This change can present challenges for leaders in terms of mobilising and engaging their teams. This situation is conducive to stagnation due to the dynamic nature of the environment. Furthermore, emotional factors must be considered. In the current era of uncertainties, such as conflicts and pandemics, this phenomenon has only increased. In times of uncertainty, the quality of listening and understanding is affected. Digital environments can also have a negative impact on the level of focus. According to *BBC News* (https://www.bbc.com/news/articles/cd9qjjj4zy5o), a lab leak is the most likely source of the COVID-19 outbreak. If the novel coronavirus was indeed developed and leaked from a laboratory, there is also a growing sense that technological advancements are posing a threat to human existence. This shift is particularly impactful on social bonds and face-to-face interactions, leading to a complex situation regarding human emotional and social needs. Digital transformation is having a significant impact on various aspects of leadership and HR, creating a need for agile, adaptable, and technologically integrated leadership.

The integration of AI and digital twin technologies has completely transformed the nature of leadership by providing data-driven, predictive, and instantly adaptive decision support systems. In today's professional environment, leadership is not solely defined by intuition and experience. The process is characterised by strategic reasoning, supported by Big Data analytics, algorithms, and scenario-based simulations. This transformation has brought about significant changes in both decision-making processes and leadership styles.

10.3 Human Resources as a Battery of Digital Culture

HR can be viewed as a key component in achieving organisational sustainability. When organisational culture integrates digital culture into its structure and HR as a fundamental component of organisational integrity, it can enhance the organisational capacity to achieve corporate goals.

Digital transformation is an ongoing process. Businesses and organisations are adopting digital technology in order to make significant changes to their core business processes, cultural structures, and customer interactions. This transformation encompasses not only the integration of technological tools but also significant changes in business processes, strategic objectives, and corporate culture. In today's rapidly changing business world, digital transformation is a key factor in achieving competitive advantage and achieving sustainable growth. This transformation enables companies to enhance their operational efficiency, improve customer experience, and differentiate themselves in the market by offering innovative products and services. Digital transformation also fosters greater flexibility and adaptability, enabling organisations to respond swiftly to evolving market conditions (PricewaterhouseCoopers (PWC) International Limited, 2025).

Infrastructure is a vital component for implementing management models and systems. One of the most critical elements of this infrastructure is the culture component. To facilitate AI-based technological transformation, it is essential to have a resilient and agile digital culture. Digital transformation, which entails the integration of digital technologies and AI into various aspects of business operations and HR, is becoming a pivotal element of conducting business and working life and is being thoroughly embraced by organisations.

We are all human, and we all make mistakes. The integration of digital technologies and AI, or the replacement of HR, will contribute to the mitigation of these errors, the reduction of the impact of human fatigue on activities and operations, the minimisation of the necessity for basic training and training repetitions as the technology develops through self-learning, and the complete elimination of these errors in specific business areas. In this respect, increasing the capacity of error reduction, prevention, and risk management creates a vital opportunity in terms of protecting resources. Furthermore, the recruitment process in HR, from application to final decision, is a costly undertaking. Workforce planning can require significant investment due to the necessity of separate production planning in each sector. The allocation of workloads in accordance with legal and regulatory obligations, as well as workforce targets and performance metrics, can result in significant financial implications, particularly for large enterprises. Even the smallest mistakes can have significant repercussions. It is important to ensure that the correct person is in the correct position, as misplaced trust can have a significant impact on the business. The implementation of AI has been a key factor in simplifying this process, optimising processes and facilitating cost savings. There is a perception of bias against AI. In order to address this issue, leaders must provide clear answers to employees' questions. Will employees be replaced by machines or robots? What are the likely consequences for employment levels? Empowering staff to develop digital literacy skills is key to their integration and adoption of digital technologies, including AI, in the workplace. Another key issue is accountability. For instance, in the event of a malfunction in digital systems, who will assume responsibility for this issue? This is a crucial management concern.

Digital culture is when technology is a vital part of the business landscape, not a mere tool. Employees need to know how to use digital tools and software to make this change. Digital transformation is most successfully achieved by fully integrating technology into business processes. Potential points of resistance should be identified by asking employees for feedback, and value should be created through pilot projects. HR professionals need to understand that digital transformation does not threaten their role. It creates opportunities to improve productivity, job security, and creative thinking. It's important for HR to see digital technology as an opportunity, not as a threat. This helps to make sure the organisation functions as well as possible.

The role of AI and automation in the digital age is a subject of significant interest. Systems such as Manus, for instance, are designed to accelerate and guarantee the precision of recruitment processes, thus enhancing productivity in white-collar roles. However, it is crucial to understand that digital transformation encompasses more than simply the use of technology; it requires a cultural and leadership environment that fosters its adoption and integration. When Manus receives CVs in a zip file, it does not merely sort candidates. Instead, it meticulously examines each CV, extracts relevant skills, and aligns them with prevailing market trends. This comprehensive approach enables the system to formulate an optimised hiring decision. Additionally, Manus has the capacity to generate its own Excel spreadsheet to present these findings (Sozcu, 2025). Digital transformation in the business world necessitates entrepreneurial risk-taking. Leaders, therefore, are capable of assuming and taking digital and technological risks. The ability to persuade and inspire is paramount for leaders when devising and applying digital transformation initiatives too. It is important for leaders to establish digital culture transformation simultaneously with their initiatives and to develop and implement their strategies to include all HR.

NLP, one of the subfields of AI, is transforming HR management. The development of digital technologies, in conjunction with the integration of organisations into leadership and HR management processes, has resulted in a paradigm shift in decision-making processes. This transformation is regarded not only as a technical innovation but also as a radical change in managerial mindsets and strategic management approaches. In today's business world, leaders are increasingly turning to AI-powered systems to inform their decision-making processes. These systems are designed to draw on large data sets, providing leaders with valuable insights and information to inform their strategic decisions (Tambe *et al.*, 2019). The digital leadership approach provides risk leaders and leader managers with the environment and/or support systems to predict employee behaviour based on past and present performance data, to perform scenario simulations in crisis situations, and to analyse team dynamics more precisely. It is imperative that they use analysis and evaluation techniques together with risk sources. Otherwise, decisions may be made based on erroneous assumptions made by the algorithm rather than on the data itself. Poor decision-making based on incorrect assumptions can have a negative impact on organisational climate, employee engagement, and individual performance, ultimately affecting corporate performance and productivity. In addition, HR departments are now leveraging advanced AI algorithms to perform a range of critical functions, including recruitment, performance evaluation, training planning, career development, and workforce analytics. These processes are often faster, more accurate, and more objective than traditional methods. These technologies facilitate processes and make significant contributions to organisational justice, increasing employee engagement and more effective human potential management (Strohmeier and Piazza, 2015). The most critical need of industries is to incorporate human-friendly, complementary technologies/relevant systems to ensure that people always feel that they are and always will be the most vital part of the process. Otherwise, people may feel threatened or that they have no place or meaning in the system. This will reduce the motivation and performance of HR to incorporate their own skills and competences into their work. The use of the term "machine next to human" as opposed to "human instead of machine" is more beneficial in terms of fostering the trust that is essential in such situations (see Figure 10.2).

Today's and perhaps tomorrow's biggest topic of discussion is that "a job done by autonomous systems without human hands and spirit will never be complete and/or good enough". Is this a psychology of reflection involving fear? This is also a topic for a separate discussion. But what is certain is that the replacement of humans by AI reduces people's motivation even as an idea, creating anxiety and tension. Keeping the motivation level at the desired level is one of the most sensitive and difficult areas of the business world. In addition, there is a risk that autonomous systems will make people passive and lazy – so much so that, for example, the fact

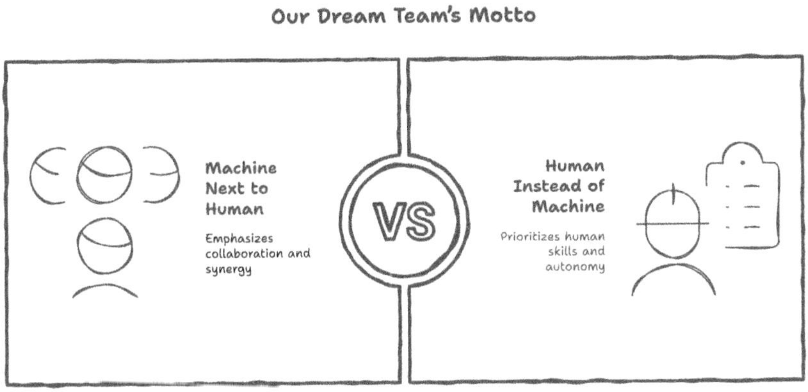

Figure 10.2. Our dream team's motto.

Source: Created using Napkin AI (https://www.napkin.ai).

that the autopilot is active at different stages of the flight and for a long time creates inertia for the pilot to take action at the required time. Similarly, leaders' unbalanced inclusion of AI in their decision-making processes can also overshadow their original thinking and idea development. Today, digital addictions such as technology addiction, smartphone addiction, and internet addiction prevent capacity utilisation both physically and cognitively, and they create physical and mental problems. Virtualisation, distancing yourself from real life, and loneliness are the feared risks of tomorrow.

At this point, it would be a good idea to add AI to our hybrid leadership model. Many experts recommend the "AI + human approval" model in aviation HR processes so that the data-driven recommendation offered by the algorithm is blended with the experience of human experts. For example, in candidate pre-selection, the algorithm reduces thousands of applications to 200, but the final interview and decision is still in the hands of human managers. This hybrid structure makes it possible to get closer to the best results, maintaining both speed and the human touch. Especially considering that the majority of accidents are caused by human error, AI tools that systematically detect and correct errors are of great value. However, *with the understanding of "let the people have the last word"*, flexibility is provided in unexpected situations (Gryncewicz *et al.*, 2023; Materna *et al.*, 2023).

The risk of misperception can also be mitigated by adhering to the principle of treating individuals not as machines, but as human beings.

> *As an AI, I significantly impact HR and leadership by boosting efficiency and providing data-driven insights for decisions. However, my limitations lie in areas requiring genuine human qualities: deep empathy, ethical judgement, truly original creativity, navigating complex social dynamics, adapting to the unpredictable, and inspiring belief. While AI enhances these fields, the unique emotional, ethical, creative, and social abilities of humans remain irreplaceable.* (Created by Authors using Gemini 2.0 Flash, Google, May 2025)

Furthermore, it is crucial to ensure that individuals, who may be experiencing feelings of isolation, fearing automation, or facing the prospect of job loss, have the assurance that there will always be a demand for human expertise and skills as machines continue to evolve. Therefore, human–machine integration is not only a digital or technical industry issue. This integration is multifaceted, especially in terms of cognitive features, ergonomics, organisational restructuring, and management function reorganisation.

Today, the most critical issue in HR performance management is the difficulty in achieving a work–life balance. NLP can help organisations understand employees' attitudes, engagement, and emotions by analysing the feedback obtained using employee sentiment analysis, NLP, and machine learning techniques in HR management. For example, the Intelligent Pulse (i-Pulse) system evaluates this feedback to inform management about employees' emotions and engagement levels and to create personalised strategies to improve work–life balance. NLP shows that it is an integrative tool in communication and contributes to management decision-making processes. For this kind of analysis to have meaningful and beneficial effects in businesses, it is important that solutions such as i-Pulse become an integral part of businesses (Garg *et al.*, 2021; Panayides, 2023). Organisational communication (message transmission and feedback) is the determinant of the performance of organisations. NLP can also support the health of organisational communication through accurate analysis and accurate and clear detection of gaps in communication.

10.4 Will Robots Take over the World? Is a Robot Leader a Realistic Possibility?

In today's business world, the field of robotics has become a major centre of attention. The increasing complexity of robots, the difficulty of maintaining safety and health, the growing population, and decreasing resources are making it necessary for robots to assist individuals and companies. At this time, it is essential to thoroughly analyse the probability of robotic leadership and the risks it will create. However, given the numerous intricacies involved in risk analysis, it is imperative for today's leaders to devise strategic alternatives for situations that cannot be fully predicted or mitigated.

The prospect of robot chefs in the workplace is an idea that many of us may find hard to imagine. It seems unlikely that any human would want to take instructions from a robot, but the ability to give instructions to robots is an aspect that could be enjoyed by humans. The concept of robots as helpers is something we are increasingly familiar with in today's society. For those who do not remember the cartoon *The Jetsons*, which was released in 1962, here is a brief overview. It was set 100 years in the future, in 2062, and many of the technologies depicted in the cartoon have either since become reality or are rapidly approaching commercial viability.

The Jetsons, the 1962 animated series that humorously portrayed the technological world of the future, did not feature a robot leader. Instead, it showcased humanoid robots such as Rosie the Robot Maid. Rosie was a robot maid who assisted the Jetsons family with domestic duties and developed a close relationship with the family members. Consistent with the series' overall theme, it depicted a world where technology simplifies life for humans.

Historically, robots have frequently been portrayed in cartoons and popular films as supportive assistants, as exemplified by characters such as Rosie the Robot Maid in *The Jetsons* or R2-D2 in *Star Wars*. However, they were never portrayed as decision-makers. In the current business environment, there is an increasing level of speculation about the potential for robot leaders to replace human leaders. However, there is a lack of clarity regarding the extent to which people will be willing to draw

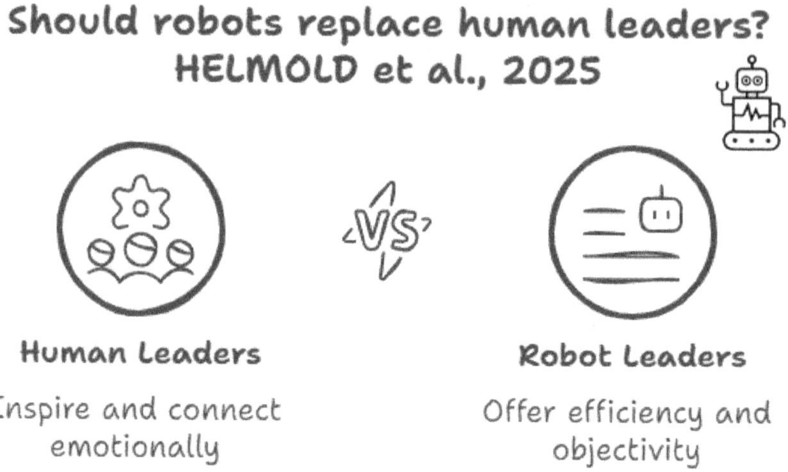

Figure 10.3. Is it possible to replace human leaders with robots?
Source: Created using Napkin AI (https://www.napkin.ai).

inspiration from robots or to accept and follow them as role models. What is your view on human leadership replacing robots? (see Figure 10.3).

In the television series *The Jetsons*, Rosie the Robot is the intelligent and articulate housekeeper responsible for household chores. A modern equivalent of Rosie is the Roomba, a vacuum-cleaning robot developed by iRobot. This can be programmed and controlled using Alexa or Google Assistant for hands-free cleaning. Robots have become a commonplace feature in our daily lives. A research robot has been developed to assist with household tasks. The robot has been named after Rosie and is equipped with four-fingered hands, a mobile base that moves in any direction, and a variety of sensors. In India, the startup Zimplistic has developed a kitchen robot that makes roti, saving home cooks a total of 100,000 hours per year. Another example is Leka, a smart robotic toy designed to support children on the autism spectrum in interacting with their environment. The developers used the Dassault Systems 3DEXPERIENCE platform and digital twin technology to track eye movements and study emotional and behavioural responses (Russell, 2023).

Changes in people's personal and professional lives naturally have an effect on how they think, behave, and communicate. This, in turn, affects their work practices and the way they interact with colleagues. In the current business environment, there is an increasing demand for competencies that are not easily replicated, such as creativity, originality, empathy, emotional intelligence, and human-centred approaches. In addition, digital proficiency and maturity levels have become comparative metrics. Therefore, it has become imperative to possess digital literacy and analytical thinking skills. It is evident that both past and current expectations have been met.

It is at this point that the complexity of the situation becomes evident. It is interesting to note that, on the one hand, the field of AI is facilitating work and enhancing productivity; on the other hand, it is creating a challenge for both individuals and organisations to adapt and maintain balance. The reaction environment presents a dual challenge: It poses risks and creates opportunities. Risks created by AI are interactive risks, with the potential to create new threats and opportunities. The environment is characterised by a yin–yang dynamic. Leaders will once again use AI to strike the right balance. How can business leaders achieve this balance?

Risk scenario: Risks of cyberattacks!

Things get easier and more productive, but will this create a situation where there are no black or grey areas? We wish it wasn't so, but every situation has positive and negative aspects. The risk management approach involves dealing with the possible negative effects of the risks and making the most of the good things that could happen as a result. Nowadays, you can see creative scenarios in TV series or films in this context. There have been many cyberattacks.

For example, Android and iOS have different operating systems and features, with the main difference being their structure. Android is an open-source system developed by Google and built on the Linux kernel. It can be used by many different manufacturers and is open to customisation. Apps are installed through the Google Play Store and run on an ad revenue model. iOS is a closed system developed by Apple and only runs on Apple devices. It has a more secure and stable structure but is closed to customisation. Apps can only be installed through the App Store. iOS maintains device performance with frequent updates (İstanbul Gelisim University IT Department, 2025). An attack on even one of these systems would have a significant impact on individual users and various stakeholders in the technology ecosystem (e.g., users, Google, the owner of Android, app developers, mobile operators, and advertising and marketing companies). In addition, if a cyberattack affects multiple systems simultaneously and causes disruption, even for a short period of time, a chain of events with potentially catastrophic social and economic consequences could be set in motion. as we have seen, even when disruptions are announced, there are problems. The potential for harm is even greater in the case of unannounced and/or unexpected and/or man-made disruptions to systems.

In a recent instance, on 28 April 2025, air traffic control systems at Newark Liberty Airport were down for about 90 seconds. This led to the departure of some controllers on "trauma leave" and delays in flights. Details of the incident were reported in the press with the information that there were no security measures and that the controllers were concerned about the problems at the airport. The FAA explained that outages can create stress and that they are working to improve the reliability of the equipment (NTV, 2025).

For this reason, one of the possibilities of AI in HR is to make recruitment, competency analysis, evaluation of candidates, and the decision to hire the best candidate for the job. However, if this system is blindly trusted or if the system is transformed by people with ulterior motives in accordance with their own interests, recruitment decisions will not be objective but can be produced as decisions in accordance with the interests of some groups. This can lead to unfair recruitment. For this reason, the control system should be included in the decision-making processes of developing and not fully mastered systems, and the outputs of the system should be controlled. People are not all the same. There are people with good intentions and people with bad intentions. It is necessary to be prepared for the threat scenarios that can be created by AI experts working with bad or ill-intentioned people. The fact that AI and digitalisation are developing at a very fast pace makes it difficult to control and develop all scenarios. While the digitalisation of systems increases their use, performance, and speed, its sudden interruption means that the systems will continue to move quickly but uncontrollably.

The capacity of AI to redefine industrial structures, transform HR markets, and influence geopolitical dynamics gives rise to significant questions regarding the future of humanity. However, by adopting a proactive and responsible approach to AI management, it is possible to mitigate potential risks, maximise benefits, and ensure that AI remains aligned with human values and interests. A range of strategies have been posited as effective mechanisms for the management of AI-related threats, including the establishment of clear ethical guidelines, the promotion of interdisciplinary collaboration, the implementation of transparency and accountability measures within AI development processes, the execution of regulatory oversight, the facilitation of public participation, and the continuous monitoring and evaluation of AI systems. The implementation of these strategies has the potential to harness the transformative capacity of AI, while concurrently establishing safeguards against the inherent risks of such technology (Farahani and Ghasemi, 2024). In order to make predictions for the future, it would be beneficial to utilise the results of AI in HR in conjunction with real-life examples from the sector as follows.

AI plays an important role in transforming airports. Contactless ID checks and AI-powered systems improve the passenger experience and optimise operational processes. Applications such as recommendation engines, chatbots, easy baggage scanning systems, and facial recognition increase passenger satisfaction and ensure safety. Furthermore, AI's data processing capabilities are making air travel safer and more efficient (TAV Technologies, 2025).

As a subsidiary of TAV Airports, TAV Technologies provides international services to more than 50 airports in more than 20 countries on four continents with its expert staff in airport IT solutions and technology products that are among the world's leading brands (TAV Airports Holding, 2025). TAV Technologies aims to improve the passenger experience by using AI in airport solutions. The AI-powered pre-order system provides personalised recommendations for passengers' purchases to increase sales. In addition, with an algorithm that predicts flight delays, airports improve the customer experience by increasing operational efficiency. Modern airports aim to provide a smoother travel experience by adopting new technologies for safety and efficiency. The integration of AI and automation in airlines is experiencing a significant transformation.

Fraport Group's AI@Fraport initiative is a digital transformation project using AI to make processes more efficient and redesign work environments. Its goal is to make Fraport fit for the future while increasing satisfaction among airlines, passengers, and staff. The initiative is organised in various areas, each with a different focus – on the administrative and operational uses of AI or addressing basic aspects such as defining ethical principles, for example. Each area strengthens the Group's digital competencies and optimises its business processes (FRAPORT, 2025). As part of the "Fraport.2030" strategy, AI@Fraport shows the group's commitment to top-notch performance worldwide. Fraport takes challenges, e.g., the shortage of qualified workers and customers' increasing expectations, very seriously and strives to meet them with creative solutions. Teamwork across departments and divisions promotes group-wide sharing of information and collaboration with partners to develop practical AI solutions. This initiative is driving innovative solutions by focusing on AI and integrating it worldwide in Fraport AG's multifaceted processes. The group is paving the way for intelligently networked airport operations in the future. Fraport has already been addressing ethics issues in connection with AI since 2021. In 2024, it became one of the first in the industry to establish worldwide principles for all AI activities (FRAPORT, 2025).

The Lufthansa Group is leveraging AI across its processes to enhance the travel experience, optimise operations, and boost sustainability. The Lufthansa Innovation Hub (LIH) is dedicated to the development of cutting-edge technologies in a variety of domains, including the optimisation of flight routes and the provision of virtual reality solutions. While CO_2 emissions are being reduced as a result of projects carried out with Google, business travel bookings can be completed in a matter of minutes via chat using an AI-supported assistant named Swifty. This system, developed with ChatGPT-4 integration, aims to combine transport, accommodation, and other travel services on a single platform in the future. Lufthansa positions AI as a transformative catalyst for the future of travel. As AI continues to evolve, its impact on the travel industry becomes increasingly apparent. From optimising travel planning to enhancing sustainability, AI is reshaping the way we experience the world (Johnson, 2024).

Recent research conducted on commercial airline companies in Portugal has demonstrated that HR departments are able to measure employee value more effectively by leveraging AI-supported data analysis. This approach has been shown to yield more successful outcomes in areas such as rotation planning, shift management, and long-term performance predictions (Pimenta de Brito, 2025).

The aviation industry is being transformed by AI in many areas, from autonomous flight technologies to air traffic management, and from passenger experience to operational efficiency. Leading organisations such as Boeing, Airbus and Amazon Prime Air are actively using AI-supported systems in flight control, navigation, and automatic landing-take-off processes. As part of the ATTOL project, Airbus has successfully tested the use of AI for the taxi, take-off, and landing operations of aircraft. AI also plays an active role in air traffic management. Projects carried out by organisations such as SESAR and NASA provide route optimisation and fuel savings by predicting traffic density. In terms of passenger experience, airlines such as KLM are offering personalised services using chatbots and virtual assistants, while airport operations are being accelerated with security solutions such as facial recognition and behavioural analysis. AI-based systems also optimise load distribution and route planning in air cargo operations, while DHL is actively using these technologies for faster and safer deliveries. In the context of pilot training, the enhancement of decision-making competencies is achieved through the utilisation of real-time simulations. Examples such as United Airlines, Delta Airlines, and IBM Watson demonstrate the environmental and operational gains of AI, such as weather forecasting, route optimisation, and reduction of carbon emissions. Heathrow Airport has implemented AI systems to optimise resource management and enhance security processes by analysing terminal density. The use of AI is key to maintaining the current level of competitiveness in the aviation industry while also establishing the foundation for the future of autonomous and sustainable flight (EASA, 2020; Airbus, 2020; SESAR, 2021; NASA, 2021; KLM, 2020).

Another example of this is Amazon Prime Air's package deliveries with its AI-powered drones. These drones are able to make safe deliveries by avoiding obstacles while analysing environmental factors and choosing the most suitable route (Amazon Prime Air, 2020). AI is revolutionising the aviation industry by enhancing passenger experience, improving operational efficiency, and increasing safety. AI-powered chatbots streamline ticket booking, flight status tracking, and customer support. Facial recognition technology accelerates security processes, while AI-driven systems optimise cargo logistics and fuel management, reducing costs and carbon emissions. Additionally, AI simulations aid in pilot training, weather prediction systems enhance flight safety, and AI-driven analytics improve airport operations, such as security and baggage handling. Leading airlines and airports, such as KLM, DHL, and Heathrow, are leveraging AI to optimise services and improve overall performance (Transportation Security Administration, 2019; DHL, 2021).

The aviation industry stands to benefit greatly from the significant solutions that AI can offer, including increased operational efficiency, optimised fuel consumption and improved safety. For instance, United Airlines employs AI algorithms to determine the most efficient flight routes by analysing weather and aircraft load, thereby saving fuel and reducing carbon emissions. Delta Airlines utilises AI to predict turbulence and adverse weather conditions, providing pilots with early warnings, thereby enhancing flight safety and comfort. In addition, Heathrow Airport is using AI to analyse passenger congestion at security passes and optimise waiting times, so that resources in the terminal are managed more efficiently. AI is transforming the airline industry, enhancing operational efficiency, reducing fuel consumption and costs, and improving passenger experience (United Airlines, 2021; Delta Airlines, 2020; Heathrow, 2019; IBM Watson, 2019).

In accordance with real sector practices in the business world, In the context of HR, AI has emerged as a significant opportunity, particularly in the areas of recruitment, competency analysis, evaluating candidates, and making optimal hiring decisions. However, when placing unquestioning trust in the system, there is a risk of it being influenced by individuals with ulterior motives, potentially resulting in decisions that are not objective but rather aligned with the interests of certain groups. This can potentially lead to unfair hiring practices. To mitigate these risks, it is essential to incorporate a control system into the decision-making process for developing AI systems rather than relying solely on the outputs of these systems. It is important to recognise that individuals have different intentions and that some may have malicious intentions. Organisations should be prepared to face potential threats from AI experts who may collaborate with malicious actors. The rapid advancements in AI and digitalisation present challenges in anticipating and managing all potential scenarios, as the digitalisation of systems enhances their utilisation, performance, and speed. However, an unexpected interruption to these systems can lead to uncontrolled and rapid changes.

The contribution of AI in HR applications is mostly observed in three main areas. First, algorithmic recruitment processes allow you to select the right candidate faster and with less bias. The NLP algorithms used in this process determine the candidates' suitability levels by analysing their resumes and motivation letters. Secondly, during the monitoring and evaluation of employee performance, AI works integrated with systems that can instantly analyse the activities and contribution levels of employees. Finally, in areas such as career development and training planning, AI statistically analyses the strengths and weaknesses of employees and proposes individualised development programmes (Pimenta de Brito, 2025). Imagine that the entire signalling system is disabled, and the communication system is disrupted. There would be terrible accidents and irreparable moral and material losses as a result. While this is a highly unlikely scenario, it is essential to consider the potential consequences and implement measures to mitigate these risks.

10.5 Decision-Making and Business Behaviour: Hybrid ADiReR Leadership

The hybrid ADiReR leadership plus AI is a contemporary leadership approach that effectively manages risks in a business environment that is variable, uncertain, complex, and full of difficulties in making sense of, with a high level of complexity and uncertainties that cannot be controlled. Hybrid leadership is a concept that takes traditional leadership approaches and enhances them. It is a paradigm that is focused on digital culture and characterised by the provision of freedom and flexibility to motivate teams, ensure collaboration, and facilitate rapid adaptation. This approach enables leaders to assume a leading role within organisations as digital leaders. This, in turn, facilitates the achievement of targeted results in terms of sustainable, creative strategies. As a result, agile, motivated, innovative, and willing HR can create too. Actually, here, the agile concept includes not only the role of a leader but also the participation and responsibility of team members. This approach is important for organisations that want to be successful and achieve sustainable results in a rapidly changing business world (Flow Coaching Institute, 2025).

Decision-making is a cognitive, social, and organisational function that allows individuals or institutions to choose among alternatives towards a specific goal. This process includes not only data-driven rational analysis but also intuition, experience, contextual evaluation, and value judgements (Kumar *et al*., 2024). AI technologies can play a supporting role in these processes and increase the decision capacity of leaders, especially in areas such as data analysis, scenario creation, and risk assessment. The contribution of AI lies in its ability to model multivariate systems and offer predictive comparisons between alternative outcomes (Balbaa, 2024). AI can enhance decision-making, automate routine tasks and streamline processes. Organisations of all types and sizes are using AI to improve operations, customer experience, and innovation. AI is driving efficiency, data-driven insights, and automation across various industries. As AI adoption increases, companies must ensure fairness, privacy, and compliance. To avoid unintended harm and build trust, businesses using AI must prioritise ethical implementation in business applications such as hiring, customer service, and financial decision-making. They can balance innovation and responsibility by proactively addressing bias, improving AI transparency, safeguarding data and complying with the law. Ethical AI enhances credibility and long-term success, not just risk management (Lamb, 2025). AI has transformed and continues to transform the way of doing business and the attitude of the individual in both social and business life by adding new dimensions. In previous generations, change and transformation used to take place over a long period of time, so people might have resisted change in an environment of slow or gradual transformation to a new situation. But today, the effect of change and transformation – of course, with the effects of AI and digitalisation – has accelerated so much that it is not possible to resist. Competencies related to the speed of adaptation have become the topic of the day.

The main topics in today's sectoral and/or academic research focus on the speed of digitalisation, digital competence (maturity) levels of businesses, and HR. In fact, in this highly accelerated transformation environment, leaders remaining unchanged and undeveloped is out of the question. Because in an environment where individuals experience sharp transformations and changes, competency transformations, business models, and working models have been influenced by AI. The level of influence of the business world from AI is increasing. In such an industry environment, leaders need to keep their competencies in use in this environment. Today, the communication style of the new generation and even the older generations like us in this new environment has changed. Our communication tools have also changed and diversified. If the instruments, perceptions, and attitudes about communication are changing, then everything is changing. The instruments that individuals use in their self-leadership are also being renewed and diversified.

AI has emerged as a transformative force, impacting various aspects of business operations, from the production line to the boardroom, and generating opportunities for growth and enhancing competitiveness (Assis, 2024). The human factor is not only a technical complement in decision systems powered by AI; it is also the bearer of the ethical framework and the guarantor of legitimacy. AI's capacity to generate decisions produces societal acceptance and managerial trust only when it works in tandem with human intuition, judgement, and responsibility (Radanliev, 2025). But as importantly, AI ethics does not only limit the limits of technological progress; it is also the basic normative framework that determines how this progress will be managed, according to which values it will be shaped and how it will gain its social legitimacy. For AI systems to operate reliably, fairly, and responsibly, ethical principles must become not only guidelines but also the standard of institutional practice (Osasona *et al.*, 2024). Hybrid ADiReR leaders can pursue a strategy to increase productivity in industries by incorporating digital technologies into the business, but a new risk is emerging here: digital fatigue and workload. To manage these factors, strategies such as training, technology utilisation, resource dependency management, and time management are among the strategies risk leaders consider.

Technological innovations resulting from digitalisation must be managed considering the human factor, which is considered as a potential source of risk. Strategies that prioritise workload management, fatigue reduction, and the promotion of human factors principles are considered essential to improve safety and operational efficiency in the aviation sector. Achieving this balance requires the integration of various approaches, including training, use of technology, and effective time management. Digitalization has transformative potential to both opeations and organizational functions with their processes. On the one hand, in an environment where the social and business world under the influence of digitalisation and AI makes people lonely, negatively affects their communication, and everything is consumed very fast, people's need for leadership is increasing in a sense. There is no hard work in business life, especially in the business world, where almost perfect solutions are produced with the effect of digitalisation and AI. But the most important problem today, as in the past, is communication. Although digitalisation and AI provide the optimum environment for communication, there will always be communication problems where there are people. Communication is the key point of digital agile leadership, like any other leadership style. If there is a problem in communication, there is a problem in everything. When there are disruptions and problems in communication, all managerial and operational processes and all corporate goals are negatively affected.

Effective communication plays a key role in any industry and their business sectors, where precise implementation of processes and procedures is vital. In business, where national, organisational, and professional cultures influence the way employees communicate, interpret meaning, value performance, and assess risk, communication is a key skill for leadership. One of the most prominent characteristics of leaders is their ability to make themselves heard, listened to, followed, and sustained by individuals and audiences. Especially in an environment that is increasingly digitalised and shaped by the external voice of AI, communication requires different methods and manoeuvres from before. The dynamics of the virtual environment are also involved. On-site applications have been replaced by remote or hybrid applications. It would not be wrong to say that as the social and business environment in which the people being led live is transforming, the human being is also transforming. At this point, leaders need to gain new competencies by accepting the differences under the influence of digital and AI.

We are in an environment where motivation is very difficult. Everything is very fast, and everything is done much better and more completely by machines and AI. People are experiencing deeper fears in the real world today than they used to experience in movies. Unmanned passenger aeroplane projects have already started. But research shows that people still do not prefer to board pilotless aeroplanes by a clear margin. And why is that?

In the "artificialised" environment, people still prefer to trust humans the most. No matter how much AI develops, can it fully replace humans? What is still most valuable are those that cannot be imitated? Core competencies are the most critical area where companies gain competitive advantage. And the most valuable core talent that cannot be imitated is HR. Each person is unique. While digitalisation is moving towards a uniform model, human originality is the power to create, the power to make mistakes, and the power to inspire and be inspired. It is difficult to say exactly and clearly where these advancements will lead, but it can be said with certainty that human beings will always need leaders and that the digitalised and AI-influenced environment increases the human need for this, rather than decreasing it.

We suggest that the powerful solution for today's transformed business world is the "Hybrid Leadership Model". This model includes agile, digital, resilient, risk-based attitude and decision-making. In a global context characterised by the influence of digitalisation, AI, and uncertainties, leaders must not only manage but also demonstrate agility, digital competence, transformative mentorship, effective perception management, and the ability to motivate. However, this transformation is not sustainable without a robust corporate risk management framework and effective risk leadership. Hybrid leaders (ADiReR) must be able to anticipate uncertainties while overseeing change and implement proactive strategies to ensure their organisations are resilient against crises and potential threats. Agility entails the ability to swiftly adapt to change and make flexible decisions, while digital leadership necessitates the effective utilisation of technology and adapting organisations to this transformation. Transformative leadership encompasses not only management but also inspiration, guidance, and contribution to the development of the team. Perception management entails setting strategic direction by accurately analysing the expectations of employees, stakeholders, and society. Conversely, a motivational leader fosters a robust organisational culture characterised by trust, commitment, and creativity. A leader with a risk management perspective is able to not only identify uncertainties as potential threats but also transform them into strategic opportunities. In this context, contemporary leaders are not merely authority figures.

10.6 Risk-Based Optimal Solution: AI and Human Leadership Cooperation

An AI-supported hybrid leadership model can increase objectivity with data-driven decisions while preserving human ethical and emotional intelligence. While AI offers alternative solutions by making analyses, the final decision should be made by humans and supervised by multiple control mechanisms. Thanks to transparent algorithms, it should be possible to understand how AI works, and the processes involved in decision-making should be optimised in view of corporate management. A participatory and inclusive process is essential for ensuring the involvement of employees and stakeholders, and AI can contribute to a fairer understanding of leadership by balancing the imbalances in human nature.

The collaboration between leaders and AI has become a critical factor in achieving agile risk management. It is necessary to adopt a leadership approach that integrates elements such as AI-supported decision-making, data analytics and automation. This integration enables leaders to make faster and more informed decisions while offering great advantages in predicting and managing risks. In today's competitive business environment, leadership has evolved into a field where agile leaders who can collaborate effectively with AI, make data-driven decisions and adapt quickly to change stand out.

The leaders who demonstrate high levels of awareness, wisdom, and compassion are better positioned to establish the foundations of effective leadership, enhance performance, and provide a more positive experience for the individuals they lead (Hougaard and Carter, 2024). The impact of AI on leadership is transforming modern management. First and foremost, AI serves as an ever-reliable assistant, swiftly analysing data and offering managers actionable recommendations. This approach ensures that decisions are made with greater clarity and that any changes can be implemented swiftly. It also assumes responsibility for routine tasks, enabling managers to focus on more significant activities such as strategic planning, developing innovative concepts and providing individual support to their employees. Additionally, while AI cannot guarantee perfection, its advanced capacity to analyse and identify patterns provides managers with significant insights into decision-making processes. However, the ability to balance technological precision and human intuition has also become a key leadership skill. Furthermore, AI has been shown to facilitate the creation of customised training and development plans by identifying the strengths and development needs of employees (Budianto *et al.*, 2025). AI-supported HR has the potential to respond to modern employee demands where individuality and continuous development are emphasised. Only time will tell how much of this potential can be utilised. The important point in terms of risk management is to remember that AI should be "supportive". Interpersonal interaction, empathy and emotional intelligence remain key elements of successful hybrid (combination of agile, risk, resilience, and digital) leadership. Finally, for the new generation of leaders, the range of risks and the environment and conditions for managing them are also changing. Whilst AI is undoubtedly a key driver of organisational transformation at the highest levels of business, it must be noted that this brings with it a new set of challenges, including data security, ethics, privacy, the fair use of systems, infrastructure building and the development of digital capacity for organisational readiness to risk. These issues must be given full consideration as they become more significant sources of risk within the risk portfolio.

Digital transformation is imperative for companies that want to remain competitive. Without a doubt, the integration of digital technologies has proven to be an essential element to improve business performance. However, little is known about how leadership styles affect the outcome of digital transformation processes. It shows how agile leaders drive successful digital transformation through rapid decision cycles, continuous engagement in digital transformation activities, and evolutionary and revolutionary practices. This makes the organisation leaner and more efficient. Agile leadership makes employees feel like they are part of the broader decision-making process. Agile leadership, which also has a positive effect on increasing the motivation of employees, makes employees feel that they are a part of the future of the company. So, agile leadership is an appropriate approach to involve members of the organisation in digital transformation projects (Rialti and Filieri, 2024). In a hybrid environment, the hybrid ADiReR leader has the capacity to manage organisations and HR in a lean way.

The hybrid leader can effectively manage organisations and HR by adopting the principles of agile, digital competencies, risk management and resilience. In order to respond quickly to changing business conditions, this leader makes agile decisions, optimises processes and makes data-driven decisions by embracing digital transformation. It establishes systems that balance remote work and physical office work, allowing teams to be productive in both environments. At the same time, by adopting a risk management-based approach, it anticipates and minimises the uncertainties and risks that the organisation may encounter. In crisis situations, it increases the sustainability and resilience of the organisation by developing alternative plans. In line with lean management principles, it creates a more efficient working environment by eliminating unnecessary processes. With a people-oriented leadership approach, it fosters a culture that enhances employee motivation and eliminates micromanagement, empowering employees to take initiative and pursue self-development. Hybrid leadership allows organisations to gain a competitive advantage by addressing digital transformation and the human factor in a balanced way.

Hybrid solutions to a hybrid environment…

In fact, traditional leadership approaches are not being lost; on the contrary, new qualities are being added to the developing and transforming technological conditions of the day. Consequently, leaders must now use multiple instruments simultaneously and rhythmically. By making faster, more accurate, more strategic, and more predictive decisions, it needs to access more people, more data, and a variety of digital tools from different platforms and deliver its answers in this diverse environment without losing its originality and without being affected by AI. Risks are increasing and diversifying at such a rapid pace that technological gaps are proving challenging to manage, and making predictions is becoming difficult. The pace of change is accelerating, and there has been negligible stabilisation. Consequently, the situation can change abruptly and without warning.

AI has become a disruptive force in many industries, and it is becoming increasingly apparent that it is influencing leadership choices. This is changing the way leaders work and think about the future by improving data-driven decision-making and changing team dynamics (Khatri, 2023). The impact of AI on the future of leadership includes improved decision-making, inclusive leadership, collaboration, adaptability, and most importantly, empowering employee development, and it is an important step in shaping effective and dynamic leadership practices in the modern world of work (Bramly, 2024). Policymakers should position systems and policies to take advantage of analytic uses of data that HRM processes generate. Natural language processing techniques can reduce the need to pre-quantify information at the data collection stage. HR managers should consider AI applications to be an enabler of better policies rather than a substitute for human decision-making (Schulker *et al.*, 2021).

10.7 Fully Authorised AI as a New Leader

In today's complex, fast-paced, and ever-changing organisations, leadership styles must also transform. Moreover, in our rapidly changing business world, characterised by the influence of digitalisation, AI, and global uncertainties, leaders must not only manage but also demonstrate agility, digital competence, transformative mentorship, effective perception management, and the ability to motivate their teams. However, this transformation is not sustainable without a robust understanding of enterprise risk management and risk leadership. Those in risk leadership positions must be able to anticipate uncertainties while managing change. They must also develop proactive strategies to make their organisations resilient against crises and potential threats.

Agility is the ability to adapt quickly to change and make flexible decisions. Digital leadership, on the other hand, is the ability to use technology effectively and adapt organisations to this transformation. Transformative leadership involves inspiring, guiding, and contributing to the development of the team, while perception management requires setting strategic direction by accurately analysing the expectations of employees, stakeholders, and society. Concurrently, a motivational leader fosters a robust corporate culture by cultivating trust, dedication, and creativity within the organisation. An effective leader who has a risk management perspective will see uncertainties not only as threats but also as opportunities to be transformed into strategic advantages. In this context, contemporary leaders are not only authority figures but also change catalysts, the cornerstones of organisations, and visionary guides to the future.

It is our professional opinion that the field of management and strategy has the potential to be impacted by AI and presents both risks and opportunities. If we choose to give AI full authority as the new leader, it is vital to conduct a risk analysis for the following uncertainties:

- Delegation and responsibility: If the leader can delegate authority to the AI but responsibility remains with them, this is a concept worth exploring.
- Social and societal decisions: It is essential that the control framework, which is designed to ensure legitimacy, representation, and democratic oversight, is clearly defined and thoroughly considered.
- Crisis risks and interactive contextual situations: Consider whether silent information processing and intuitive decision-making will be possible. Intuitive decision-making is a core strength and competence of effective leaders.
- In the context of teaming and employee engagement with AI, it is essential to ensure that human autonomy and psychological safety are not compromised. Failure to address these issues could lead to additional risks for HR.
- Values, social sensitivities, and dilemmas: There appears to be a lack of ethical prioritisation and a failure to consider norms.
- Ensuring legal accountability and oversight is a multi-criteria decision-making problem in itself.
- Decision-making processes: It is important to note that ensuring effective decision-making by incorporating institutional trust, ethical transparency, and feedback mechanisms may present certain risks.

AI has become an integral part of decision-making across industries, from healthcare and finance to justice and hiring. It processes data, identifies patterns, and makes decisions with speed and accuracy that often surpass human capabilities. However, integrating AI into critical decision-making processes raises concerns about the role of human factors in shaping algorithmic outcomes. Human factors influence the design, implementation, and evaluation of AI systems. Human decisions can inadvertently introduce biases and errors that undermine automated decisions. These biases may be reflective of inequalities or unintended cognitive tendencies leading to outcomes that perpetuate discrimination or reinforce societal stereotypes (Charkra *et al.*, 2024).

Human biases, whether cognitive or societal, have a significant impact on the fairness and accuracy of AI systems. As AI becomes increasingly integrated into decision-making processes across various sectors, it is crucial to mitigate bias and error to ensure equitable outcomes for all. This necessitates continuous research, the creation of more advanced mitigation strategies, and the implementation of inclusive and ethical AI practices. By encouraging collaboration between researchers, developers, policymakers, and the public, we can ensure the responsible development and deployment of AI technologies that serve the greater good of society (Charkra *et al.*, 2024). In this regard, organisations maintaining emphasis on HR while collaborating closely with AI are vital.

In light of recent developments,
are we ready to adopt fully authorised
AI as a new leader in our lives?!..

Or it may be that we would prefer to have
AI as an assistant rather than as a leader,
as shown in Figure 10.4.

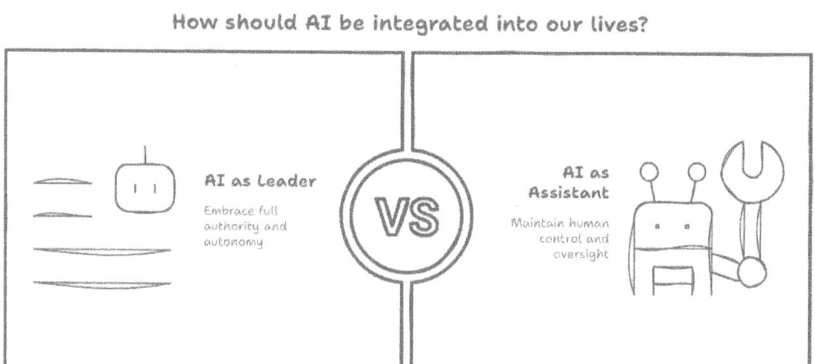

Figure 10.4. How should we integrate AI into our lives?

Source: Created using Napkin AI (https://www.napkin.ai).

AI will inevitably become an integral part of leadership decision-making and HR processes and systems. It is crucial to emphasise that AI should be structured as a supportive system under human control and/or supervision, rather than as a leader or a centralised decision authority. There is an increasing need for a balance between AI and ethical leadership. The development of technological capacity must be balanced with the preservation of human responsibility. Achieving this integration is essential for the successful fulfilment of industrial goals, while also ensuring societal and institutional trust and ethics are maintained. Maintaining this balance is crucial, as any disturbance will result in the emergence of risks that must be effectively managed. At present, a balance has not yet been achieved. This uncertainty is likely to lead to fear, anxiety, and a decline in motivation among HR, consequently negatively impacting social peace.

Bibliography

Airbus (2020) *Autonomous Flight Technologies: The future of aviation*. Toulouse: Airbus.

Amazon Prime Air (2020) *AI-Powered Drones for Delivery*. Seattle: Amazon Prime Air.

Assis, A.P. (2024) 'AI Is Changing the Shape of Leadership–How Can Business Leaders Prepare?', *World Economic Forum*, 10 May. Available at: https://www.weforum.org/stories/2024/05/ai-is-changing-the-shape-of-leadership-how-can-business-leaders-prepare/ (Accessed: 10 May 2024).

Balbaa, M. *et al.* (2024) 'The Impact of Artificial Intelligence in Decision Making: A Comprehensive Review', *EPRA International Journal of Economics, Business and Management Studies*, 11(2), 15747. doi: 10.36713/epra15747.

Bærøe, K., Miyata-Sturm, A. and Henden, E. (2020) 'How to Achieve Trustworthy Artificial Intelligence for Health', *Bulletin of the World Health Organization*, pp. 257–262. doi: 10.2471/BLT.19.237289.

Bennett, N. and Lemoine, G.J. (2014) 'What VUCA Really Means for You', *Harvard Business Review*, 12 March. Available at: https://hbr.org/2014/01/what-vuca-really-means-for-you (Accessed: 12 March 2025).

Bramly, F. (2024) '5 Ways AI is Impacting Leadership Development', *Acacia Learning*, 22 March. Available at: https://acacialearning.com/blog/management-and-leadership/5-ways-ai-is-impacting-leadership-development/ (Accessed: 22 March 2024).

Brown, A. (2024) 'The Impact of AI on the Past, Present, and Future of Leadership - Part 1', *Oxford Group*, 18 April. Available at: https://www.oxford-group.com/insights/the-impact-of-ai-on-leadership/ (Accessed: 18 April 2024).

Budianto, S., Rahadian, D. and Yunita, I. (2025) 'The Emerging Landscape of AI-Powered Leadership: Transforming Roles and Organizations', *Journal of Lifestyle and SDGs Review*, 5(2). doi: 10.47172/2965-730X.SDGsReview.v5.n02.pe04139.

Charkra, S., Dey, S. and John, A. (2024) 'The Human Factor in AI Decision-Making: Mitigating Bias and Error'. Available at: https://www.researchgate.net/publication/387428677_The_Human_Factor_in_AI_Decision-Making_Mitigating_Bias_and_Error/citation/download (Accessed: 31 December 2024).

DHL (2021) *AI-Powered Logistics in Air Cargo Operations*. Bonn: DHL Global Forwarding.

EASA (2020) *Artificial Intelligence Roadmap: A Human-Centric Approach to AI in Aviation*. Cologne: EASA.

Farahani, M.S. and Ghasemi, G. (2024) 'Will Artificial Intelligence Threaten Humanity?', *Sustainable Economies*, 2(2). doi: 10.62617/se.v2i2.65.

Flow Coaching Institute (2025) 'Leadership is Being Redefined: Being a Leader in the BANI World with Agile Leadership', 19 March. Available at: https://in-flowtr.com/urun/liderlik-yeniden-tanimlaniyor-cevik-liderlikle-bani-dunyasinda-lider-olmak/ (Accessed: 19 March 2025).

Fraport (2025) 'AI@Fraport Fraport Group', 5 September. Available at: https://www.fraport.com/en/our-group/digitalization/ai-fraport.html (Accessed: 5 September 2025).

Galinsky, A.D. (2025) 'What Sets Inspirational Leaders Apart', *Harvard Business Review*, March. Available at: https://hbr.org/2025/03/what-sets-inspirational-leaders-apart (Accessed: 30 March 2025).

Garg, R. *et al*. (2021) 'i-Pulse: A NLP Based Novel Approach for Employee Engagement in Logistics Organization', *International Journal of Information Management Data Insights*, 1. doi: 10.1016/j.jjimei.2021.100011.

Gryncewicz, W., Zygała, R. and Pilch, A. (2023) 'AI in HRM: Case Study Analysis. Preliminary Research', *Procedia Computer Science*, pp. 2351–2360. doi: 10.1016/j.procs.2023.10.226.

Hougaard, R. and Carter, J. (2024) 'How AI Can Make Us Better Leaders', *Potential Project*, April. Available at: https://www.potentialproject.com/insights/how-ai-can-make-us-better-leaders (Accessed: 30 April 2024).

Istanbul Gelisim University (2025) 'What are the Differences Between Android and iOS?', *Directorate of Information Technologies News*, April. Available at: https://bidb.gelisim.edu.tr/en/administrative-news-iosvsandroid (Accessed: 30 April 2025).

Johnson, M. (2024) 'AI in Aviation: Smarter Travel', *Lufthansa*, 19 April. Available at: https://www.lufthansa.com/in/en/articles/explore-flying/time-travel-ki-aviation (Accessed: 19 April 2024).

Khatri, R. (2023) 'The Impact of Artificial Intelligence on Leadership', *Open Growth*, 22 July. Available at: https://www.blogs.opengrowth.com/the-impact-of-artificial-intelligence-on-leadership (Accessed: 22 July 2023).

King, T.C. *et al.* (2020) 'Artificial Intelligence Crime: An Interdisciplinary Analysis of Foreseeable Threats and Solutions', *Science and Engineering Ethics*, 26, pp. 89–120. doi: 10.1007/s11948-018-00081-0.

KLM (2020) *How KLM uses Artificial Intelligence to improve customer experience.* Amsterdam: KLM Royal Dutch Airlines.

Kumar, B.R., Madhuri, A. and Bathina, S. (2024) 'The Role of Artificial Intelligence in Decision-Making Processes', *African Journal of Biological Sciences*, 6(6), pp. 6344–6362. doi: 10.33472/AFJBS.6.6.2024.6344-6362.

Kuzmanko, J. and Vrbová, L. (2024) 'AI vs Humans in Strategic Decision-Making: A Systematic Review', *SSRN*. doi: 10.2139/ssrn.5144866. Available at: https://ssrn.com/abstract=5144866 (Accessed: 30 April 2024).

Lamb, A. (2025) 'AI and Business Ethics: Balancing Efficiency with Responsibility', *Business Coaching 4 Leaf Performance*, 19 March. Available at: https://www.4leafperformance.com/ai-and-business-ethics-balancing-efficiency-with-responsibility/ (Accessed: 19 March 2025).

Maddula, S.S. (2018) 'The Impact of AI and Reciprocal Symmetry on Organizational Culture and Leadership in the Digital Economy', *Engineering International*, 6(2), pp. 201–210. doi: 10.18034/ei.v6i2.703.

Materna, M. *et al.* (2023) 'The Influence of Human Factor on Aviation Accidents in Slovakia through HFACS Framework: A Comprehensive Study', *Transportation Research Procedia*, 75, pp. 173–182. doi: 10.1016/j.trpro.2023.12.020.

NASA (2021) *AI in air traffic management: Enhancing flight safety and efficiency*. Washington: NASA.

NTV (2025) 'Chaos in the Air: Tower "Blinded" for 90 Seconds, Connection with Airplanes Lost'. Available at: https://www.ntv.com.tr/dunya/havada-kaos-kule-90-saniye-boyunca-kor-oldu-ucaklarla-baglanti-koptu,zSMeFuKJzE-G1vLBF1tmwA (Accessed: 4 December 2025).

Osasona, F. *et al.* (2024) 'Reviewing the Ethical Implications of AI in Decision Making Processes', *International Journal of Management & Entrepreneurship Research*, 6(2), pp. 322–335. doi: 10.51594/ijmer.v6i2.773.

Panayides, J. (2023) 'The Role of Natural Language Processing in Employee Sentiment Analysis', *myHRfuture*, 4 May. Available at: https://www.myhrfuture.com/blog/the-role-of-natural-language-processing-in-employee-sentiment-analysis (Accessed: 4 May 2023).

Pimenta de Brito, A. (2025) *HR Analytics in the Commercial Airline Sector in Portugal: A Mixed Method/Case Study Analysis*. PhD Thesis. Lisbon: Iscte - Instituto Universitário de Lisboa.

PricewaterhouseCoopers (2025) *Digital Transformation: Business of the Future*. Available at: https://www.pwc.com.tr/dijital-donusum (Accessed: 4 December 2025).

Rialti, R. and Filieri, R. (2024) 'Leaders, Let's Get Agile! Observing Agile Leadership in Successful Digital Transformation Projects', *Business Horizons*, 67(4), pp. 439–452.

Russell, M. (2023) '10 Things the Jetsons Got Right', *Dassault Systèmes*, 19 March. Available at: https://blog.3ds.com/topics/company-news/10-things-the-jetsons-got-right/ (Accessed: 19 March 2023).

Schulker, D. *et al.* (2021) *Can Artificial Intelligence Help Improve Air Force Talent Management? An Exploratory Application.* Santa Monica: RAND Corporation.

SESAR (2021) *Artificial Intelligence in Air Traffic Management.* Brussels: SESAR Joint Undertaking.

Sozcu (2025) 'Teknoloji nereye gidiyor? İnsan komutu olmadan tek başına çalışan yapay zeka', 11 March. Available at: https://www.sozcu.com.tr/cin-den-dunyada-bir-ilk-insan-komutu-olmadan-tek-basina-calisan-yapay-zeka-p149663 (Accessed: 11 March 2025).

Strohmeier, S. and Piazza, F. (2015) 'Artificial Intelligence Techniques in Human Resource Management — A Conceptual Exploration' in Kahraman, C. (ed.) *Intelligent Decision Making in Quality Management.* Cham: Springer, pp. 149–172. doi: 10.1007/978-3-319-17906-3_7.

Tambe, P., Cappelli, P. and Yakubovich, V. (2019) 'Artificial Intelligence in Human Resources Management: Challenges and a Path Forward', *California Management Review*, pp. 15–42. doi: 10.1177/0008125619867910.

TAV Airports Holding (2025) 'TAV Technologies'. Available at: https://tavairports.com/hakkimizda/hizmetler/tav-technologies (Accessed: 30 April 2025).

TAV Technologies (2025) 'AI in Aviation Industry', 9 May. Available at: https://tavtechnologies.aero/en-EN/review/pages/artificial-intelligence-in-aviation-industry (Accessed: 9 May 2025).

Transportation Security Administration (2019) *AI in Aviation Security: Enhancing Safety and Efficiency*. Washington: TSA.

Chapter 11

AI in the Health Sector

11.1 Global Healthcare Sector

AI is rapidly transforming the healthcare sector, offering solutions in areas such as diagnostics, treatment, and administration. AI algorithms can analyse medical imaging data to assist in diagnoses, accelerate drug discovery, and automate administrative tasks. These advancements lead to improved patient outcomes, reduced costs, and increased efficiency.

The global healthcare sector is being revolutionised by a range of innovations and technologies, unlocking benefits to patient care, health diagnostics, and operational efficiency, as well as research, development, and innovation.

The healthcare services market size is expected to see steady growth over the next few years. It will grow to US$11222.08 billion in 2029 at a compound annual growth rate (CAGR) of 4.9 percent (Global Healthcare Report, 2025). The growth during the forecast period can be attributed to telehealth and digital health, healthcare workforce shortages, health equity initiatives, value-based care, and potential shifts in reimbursement models. Major trends in the forecast period include telemedicine integration, patient-centred care, preventive care and wellness, personalised medicine, and healthcare consolidation.

The key benefits comprise the following:

Diagnostic Assistance

Medical imaging analysis
AI algorithms can analyse X-rays, MRIs, and CT scans to detect anomalies, tumours, and fractures more accurately and quickly than human doctors. For example, AI can help identify early signs of stroke in brain scans, according to the World Economic Forum.

Early disease detection
AI models can analyse patient data to identify early signs of diseases before symptoms appear. For instance, an AI model can detect certain diseases before a patient is even aware of any symptoms, according to AstraZeneca.

Drug Discovery and Development

Target identification
AI can analyse vast datasets to identify potential drug targets and optimise the design of medicinal products.

Predictive modelling
AI can help predict the efficacy of potential drugs and optimise dosing regimens.

Clinical Trial Optimisation

AI can assist with patient stratification, digital twins, and trial simulations to improve the efficiency and effectiveness of clinical trials.

Treatment Plans

Personalised medicine
AI can analyse patient data to develop personalised treatment plans tailored to individual needs and genetic makeup.

Precision diagnostics
AI can be used to identify patients at risk of specific diseases and to optimise treatment strategies.

Virtual Health Assistants

Patient education
AI-powered virtual assistants can provide patients with information and support, answering queries and offering guidance on their health conditions.

Appointment scheduling and reminders
These assistants can automate tasks such as scheduling appointments and sending reminders to patients.

Administrative Efficiency

Data entry and claims processing
AI can automate routine tasks such as data entry, claims processing and appointment scheduling, freeing up healthcare professionals' time.

Electronic health record (EHR) management
AI can assist with finding and retrieving patient records efficiently.

Robot-Assisted Surgery

Enhanced precision
AI-powered robots can perform surgeries with greater precision and accuracy than traditional methods.

Reduced surgeon fatigue
AI can assist surgeons with repetitive tasks, reducing fatigue and improving surgical outcomes.

11.2 Digitalisation and AI Applications in Healthcare

The global healthcare sector is being revolutionised by a range of technologies, unlocking benefits to patient care, health diagnostics, and operational efficiency, as well as research, development, and innovation. Experts from Nextcontinent and three of its member firms outline the following nine of those technologies that herald much promise for the coming years:

- **Telemedicine:** With the advent of digital technology, telemedicine is becoming an increasingly popular healthcare delivery model, enabling patients to receive medical care from the comfort of their own homes.
- **Artificial Intelligence and Machine Learning (ML):** AI and ML are expected to play a major role in the healthcare industry in the coming years, improving patient outcomes and reducing healthcare costs.
- **Personalised Medicine:** Personalised medicine is rapidly becoming a mainstream approach to healthcare, using genetic information to tailor medical treatments to individual patients.

- **Increased Adoption of EHRs:** EHRs are becoming increasingly prevalent in healthcare organisations, providing a more complete and accurate view of a patient's medical history.
- Healthcare providers are partnering with tech companies to develop AI tools that can better predict clinical outcomes, enhance radiological imaging, and optimise sleep monitoring.
- AI has the potential to transform healthcare by optimising both administrative functions and care delivery. It will have financial and non-financial benefits, such as improved care quality, enhanced patient experience, and greater clinician satisfaction. Private providers may gain the greatest benefits from optimisation in care, claims, and provider relationship management.
- Predictive AI could forecast patient volumes and help hospitals adjust staffing and resources by predicting future resource needs, analysing detailed data, and identifying high-impact patterns and trends.
- AI is rapidly becoming a competitive necessity in the healthcare sector. Yet, many organisations are still in the stage of understanding what AI can mean for them. Deloitte created the AI Dossier to provide leaders in different industries with summaries of key issues and opportunities, and how AI can help achieve them.

11.3 Electronic Health Records

EHRs (Figure 11.1) are crucial to the digitalisation and information dissemination of the healthcare industry. Now that around 80 percent of medical practices use EHR, the next step is to utilise AI to interpret the records and provide new information to physicians. One application uses natural language processing (NLP) to produce more concise reports, limiting variation between medical terms by matching similar ones. For example, the terms "heart attack" and "myocardial infarction" mean the same things, but physicians may use one over the other based on personal preferences. NLP algorithms consolidate these differences so that larger datasets can be analysed. Another use of NLP identifies phrases that are redundant due to repetition in a physician's notes and retains the relevant information to make it easier to read. Other applications use concept processing to analyse the information entered by the current patient's doctor to present similar cases and help the physician remember to include all relevant details.

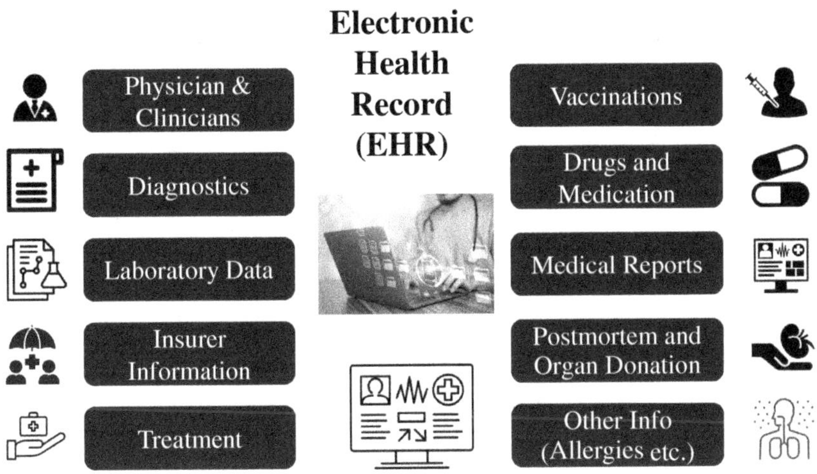

Figure 11.1. Electronic health record (EHR).

The figure shows the elements of the EHR, which can include physician names, vaccinations, diagnostics, drugs or medication, laboratory data, medical reports, insurer information, postmortem and organ donations, treatments, and other information such as allergies.

Beyond making content edits to an EHR, there are AI algorithms that evaluate an individual patient's record and predict their risk for a disease based on their previous information and family history. One generally used algorithm is a rule-based system that makes decisions similarly to how humans use flow charts. This system takes in large amounts of data and creates a set of rules that connect specific observations to concluded diagnoses. Thus, the algorithm can take in a new patient's data and try to predict the likelihood that they will have a certain condition or disease. Since the algorithms can evaluate a patient's information based on collective data, they can identify any outstanding issues to bring to a physician's attention and save time. One study conducted by the Centerstone Research Institute found that predictive modelling of EHR data has achieved 70–72 percent accuracy in predicting individualised treatment response. These methods are helpful due to the fact that the amount of online health records doubles every five years. Physicians do not have the bandwidth to process all these data manually, so AI can leverage them to assist physicians in treating their patients (Figure 11.2).

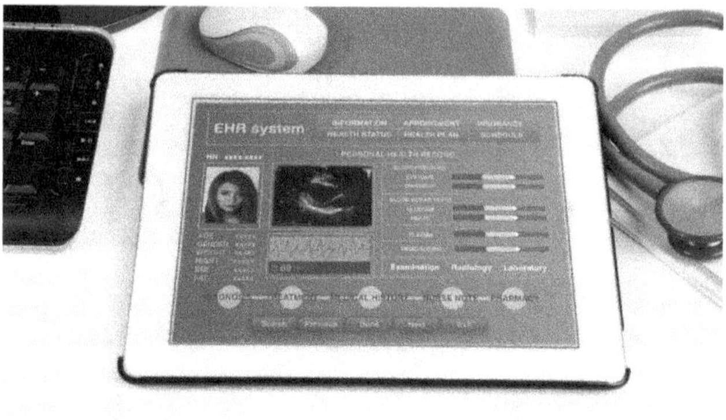

Figure 11.2. Electronic health record (EHR): Example.

Source: https://prohealthware.com/de/differences-between-electronic-medical-records-emr-and-electronic-health-records-ehr.

11.4 Primary, Secondary, and Tertiary Prevention in Healthcare

Prevention includes a wide range of activities, known as "interventions", aimed at reducing risks or threats to health. You may have heard researchers and health experts talk about three categories of prevention: primary, secondary, and tertiary, as shown in Figure 11.3 (WHO, 2024). Disease prevention, understood as specific, population-based and individual-based interventions for primary and secondary (early detection) prevention, aiming to minimise the burden of diseases and associated risk factors. Primary prevention refers to actions aimed at avoiding the manifestation of a disease (including measures to improve health through changing the impact of social and economic determinants on health; the provision of information on behavioural and medical health risks, alongside consultation and measures to decrease them at the personal and community level; nutritional and food supplementation; oral and dental hygiene education; and clinical preventive services, such as immunisation and vaccination of children, adults, and the elderly, as well as vaccination or post-exposure prophylaxis for people exposed to a communicable disease). Secondary prevention deals with early detection, which improves the chances of positive health outcomes (this comprises activities such as evidence-based screening programmes for the early detection of diseases or for the prevention of congenital malformations; and preventive drug therapies of proven effectiveness when administered

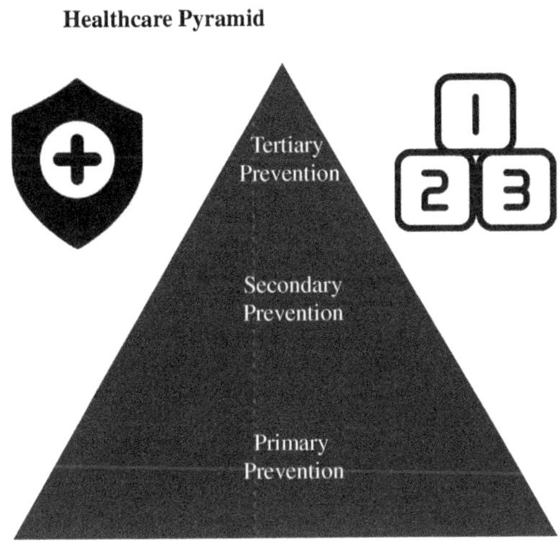

Figure 11.3. Primary, secondary, and tertiary prevention.

at an early stage of the disease). It should be noted that while primary prevention activities may be implemented independently of capacity-building in other healthcare services, this is not the case for secondary prevention. Screening and early detection are of limited value (and may even be detrimental to the patient) if abnormalities cannot be promptly corrected or treated through services from other parts of the healthcare system. Moreover, a good system of primary healthcare with a registered population facilitates the optimal organisation and delivery of accessible population-based screening programmes and should be vigorously promoted.

Tertiary prevention aims to mitigate the impact of an ongoing illness or injury that has lasting effects. This is achieved by assisting individuals to manage long-term, often complex health problems and injuries (e.g., chronic diseases, permanent impairments) in order to improve their ability to function, quality of life, and life expectancy as much as possible. Examples include:

- cardiac or stroke rehabilitation programmes and chronic disease management programmes (e.g., for diabetes, arthritis);
- support groups that allow members to share strategies for living well;
- vocational rehabilitation programmes to retrain workers for new jobs when they have recovered as much as possible.

Secondary prevention aims to reduce the impact of a disease or injury that has already occurred. This is achieved by detecting and treating disease or injury as soon as possible to halt or slow its progression, encouraging personal strategies to prevent reinjury or recurrence, and implementing programmes to return people to their original health and function to prevent long-term issues. Examples include:

- regular exams and screening tests to detect disease in its earliest stages (e.g., mammograms);
- daily, low-dose aspirins and/or diet and exercise programmes to prevent further heart attacks or strokes;
- suitably modified work so injured or ill workers can return safely to their jobs.

Primary prevention aims to prevent disease or injury before it ever occurs. This is achieved by preventing exposure to hazards causing disease or injury, altering unhealthy or unsafe behaviours that can lead to disease or injury, and increasing resistance to disease or injury should exposure occur. Examples include:

- legislation and enforcement to ban or control the use of hazardous products;
- safe healthy practices (e.g., use of seatbelts and bike helmets);
- education about healthy and safe habits (e.g., eating well, exercising, not smoking);
- immunisation against infectious diseases and viruses.

11.5 Telemedicine

The expansion of telemedicine, i.e., the remote treatment of patients, has demonstrated the rise of potential AI applications. AI can assist in caring for patients remotely by monitoring their information through sensors. A wearable device may allow for constant monitoring of a patient and the ability to detect changes that may be less readily discernible by humans. The information can be compared with other data that have already been collected using AI algorithms that alert physicians to any issues. Another application of AI is chatbot therapy. However, some researchers charge that the reliance on chatbots for mental health does not offer the

Figure 11.4. Telemedicine.

reciprocity and accountability of care that should exist in the relationship between the consumer of mental healthcare and the care provider (be it a chatbot or psychologist). Since the average lifespan has risen due to a longer life expectancy, AI could be useful in helping to care for older populations. Tools such as environmental and personal sensors can identify a person's regular activities and alert a caretaker if a behaviour or a measured vital is abnormal. Although the technology is useful, there are also discussions about the limitations of monitoring in order to respect a person's privacy, given that there are technologies designed to map out home layouts and detect human interactions. Figure 11.4 shows how telemedicine can help with diseases or diagnostics.

11.6 Machine Learning in Healthcare

The healthcare sector has long been an early adopter of technological advances. These days, ML as a subset of AI plays a key role in many health innovations, including the development of new medical procedures, the handling of patient records, and the treatment of chronic diseases (Panesar, 2021). The healthcare industry has been compiling increasingly larger data sets, often organising this information in EHRs as unstructured data. With the help of NLP, ML rearranges this data into more structured sets that healthcare professionals can quickly glean actionable insights from. Machine learning has also impacted drug discovery and development for pharmaceutical companies. The technology has already supported central nervous system clinical trials, and drugmakers hope ML will predict how patients will respond to various drugs and identify which patients stand the greatest chance of benefiting from them. In addition, telemedicine has reaped the rewards of ML evolution, as some ML companies are studying how to organise and deliver patient information to doctors during telemedicine sessions, as well as capture information during virtual visits to streamline workflows.

11.7 Drug Interaction through AI, ML, and Deep Learning

AI, ML, and deep learning (DL) innovations are used in the acceleration of traditional drug design approaches (Heifeitz, 2022). To fully capitalise on these technological advances, it is essential to systematically harness data from diverse sources and leverage digital technologies and advanced analytics to enable data-driven decisions. Data science stands at a unique moment of opportunity to lead such a transformative change (Yang, 2022). ML plays a significant role here. AI innovation has a high priority in drug design through the enhancement of ML approaches and the collection of pharmacological data. AI does not rely on any hypothetical improvements, but it has more substance in transforming medical information into studies such as reusable methods. In general, there are different approaches considered in the context of ML, including random forest, naive Bayesian classification (NBC), multiple linear regression (MLR), logistic regression (LR), linear discriminant analysis (LDA), probabilistic neural networks (PNNs), multi-layer perceptron (MLP), and support vector machines (SVMs) (Dara *et al.*, 2022). Additionally, improvements in NLP led to the development of algorithms to identify drug–drug interactions in medical literature. Drug–drug interactions pose a threat to those taking multiple medications simultaneously, and the danger increases with the number of medications being taken (Brown, 2020). To address the difficulty of tracking all known or suspected drug–drug interactions, ML algorithms have been created to extract information on interacting drugs and their possible effects from the medical literature. Efforts were consolidated in 2013 in the DDI-Extraction Challenge, in which a team of researchers at Carlos III University assembled a corpus of literature on drug–drug interactions to form a standardised test for such algorithms. Competitors were tested on their ability to accurately determine, from the text, which drugs were shown to interact and what the characteristics of their interactions were. Researchers continue to use this corpus to standardise the measurement of the effectiveness of their algorithms.

11.8 Robot-Assisted Medicine

Robot-assisted medicine, also known as robotic surgery or computer-aided surgery, involves using robotic systems and software to perform medical procedures with enhanced precision and control. This technology assists surgeons in minimally invasive procedures, offering advantages such as smaller incisions, reduced blood loss, and faster recovery times (EithHealth, 2024).

There are several application areas:

Robot-Assisted Surgery: Robots can be used to manipulate surgical instruments with greater dexterity and accuracy, allowing surgeons to perform procedures through smaller incisions, which can lead to less pain and faster recovery for patients.

Radiotherapy: Robots can be used to precisely position patients and radiation sources during radiotherapy, improving treatment accuracy.

Diagnostics: Robotic systems can be integrated into diagnostic procedures, offering healthcare professionals high-precision, safe preoperative planning and partially automated examinations.

Rehabilitation: Robots can aid in rehabilitation, helping patients regain function and strength after injuries or illnesses, for example, by facilitating movement in upper limbs or providing support for balance.

Monitoring and Therapeutics: Robots can be used for customised and frequent monitoring of patients with chronic diseases, and they can even be programmed to deliver certain treatments or provide social engagement for elderly patients.

Examples of robot-assisted systems are listed as follows:

Da Vinci System: A well-known robotic surgical system used in various surgical disciplines, such as urology and gynaecology.

Accuray's CyberKnife: A robotic radiation treatment system providing sub-millimetre accuracy.

PARO: A therapeutic robot capable of responding to touch, developing a personalised personality, and offering companionship and emotional support for patients.

Somnox Sleep Robot: A robot designed to improve sleep quality by mimicking a calming breathing pattern.

Burt Robot (Barrett Technology): Used for rehabilitation therapy to support patients with weakness or paralysis in their hands and arms.

Benefits of Robot-Assisted Medicine

Increased precision and accuracy
Robots can perform complex procedures with greater accuracy than traditional surgical methods.

Enhanced visualisation
Surgeons have better visualisation of the operating field, leading to more precise surgery.

Minimally invasive procedures
Smaller incisions mean less pain and faster recovery for patients.

Improved patient outcomes
Robot-assisted surgery and other applications can lead to improved patient outcomes, such as reduced complications and faster recovery times.

Reduced workload for healthcare staff
Robots can automate certain tasks, freeing up staff to focus on other aspects of patient care.

New treatment options
Robots can facilitate the development of innovative surgical techniques and therapies.

Bibliography

Brown, N. (2020) *Artificial Intelligence in Drug Discovery*. London: Royal Society of Chemistry.

Dara, S. *et al*. (2022) 'Machine Learning in Drug Discovery: A Review', *Artificial Intelligence Review*, 55, pp. 1947–1999.

EitHealth (2024) 'Robots in Medicine: The Robotics Start-Ups Making a Difference in Healthcare'. Available at: https://eithealth.eu/news-article/robots-in-medicine-the-robotics-start-ups-making-a-difference-in-healthcare/ (Accessed: 15 May 2025).

Global Healthcare Report (2025) 'Healthcare Service Global Market Report'. Available at: https://www.thebusinessresearchcompany.com/report/health-care-service-global-market-report (Accessed: 14 March 2025).

Heifetz, A. (2022) *Artificial Intelligence in Drug Design*. Heidelberg: Springer.

Panesar, A. (2021) *Machine Learning and AI for Healthcare: Big Data for Improved Health Outcomes*. Heidelberg: Springer.

Prohealthware (2025) 'Differences between Electronic Medical Records (EMR) and Electronic Health Records (EHR)'. Available at: https://prohealthware.com/de/differences-between-electronic-medical-records-emr-and-electronic-health-records-ehr/ (Accessed: 30 April 2025).

Suresh Mahajan, S.P. (2021) *Artificial Intelligence in Healthcare: AI, Machine Learning, and Deep and Intelligent Medicine Simplified for Everyone*. Delhi: MedMantra, LLC.

WHO (2024) 'Health Promotion and Disease Prevention Through Population-Based Interventions, Including Action to Address Social Determinants and Health Inequity'. Available at: https://www.emro.who.int/about-who/public-health-functions/health-promotion-disease-prevention.html (Accessed: 14 January 2024).

Yadav, D.K. and Gulati, A. (2023) *Artificial Intelligence and Machine Learning in Healthcare*. Wiesbaden: Springer.

Yang, H. (2022) *Data Science, AI, and Machine Learning in Drug Development*. Boca Raton: CRC Press.

Chapter 12

Data-Driven Decision-Making and Algorithmic Decision-Making

Decisions are multi-layered processes and can be made by or supported by AI (Zhang and Gao, 2025). There are several new developments in multi-level decision-making theory, techniques, and methods in both modelling and solution issues. The main concepts in AI are data-driven decision-making (DDDM) and algorithmic decision-making (ADM) (Pezotta *et al.*, 2024). There is a vast compilation of practical contributions in industry and service sectors, specifically focusing on DDDM. Practical examples show that companies that employ DDDM are more productive, have a higher market value, and deliver higher returns for their shareholders. In this chapter, the reader will discover the history, theory, and practice of DDDM, learning how organisations and individual managers alike can utilise its methods to avoid cognitive biases and improve confidence in their decisions. It argues that value does not come from data but from acting on data (Bang, 2025). DDDM enables managers to explore how data can transform decision-making processes, enabling evidence-based selection, prioritisation, and resource allocation (Stenhardt, 2024).

Figure 12.1 shows that DDDM can be implemented with partial independence of the AI or completely autonomously. It is important that the process of making decisions is transparent and ethically sound (Dignum, 2019). The figure depicts that there are options between fully autonomous or human-controlled decisions.

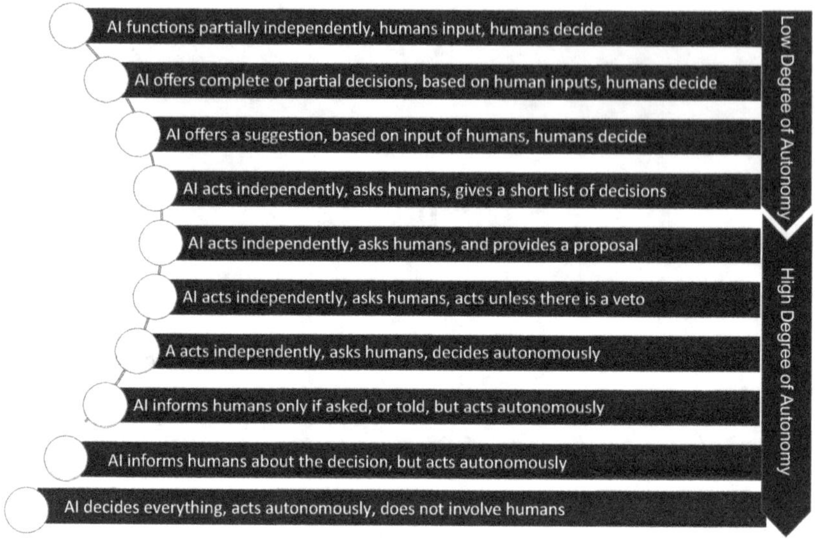

Figure 12.1. Data-driven decision-making (DDDM).

12.1 Data-Driven Decision-Making

DDDM is an approach that emphasises using data and analysis rather than intuition to inform business decisions. It involves leveraging data sources such as customer feedback, market trends, and financial data to guide the decision-making process. By collecting, analysing, and interpreting data, organisations can make better decisions that more closely align with business goals and objectives.

12.2 Benefits of DDDM

Organisations that adopt a data-driven culture see benefits such as improved customer satisfaction, better strategic planning, and enhanced outcomes.

12.2.1 Guard against bias

Data-driven decisions minimise personal bias and safeguard objectivity. Many enterprises implemented various debiasing techniques as part of their decision-making processes across the value chain. The organisation established programmes to raise awareness of cognitive biases among its leadership and employees. These strategies protect diverse perspectives and ensure that decisions are not influenced by hierarchical or confirmation biases.

12.2.2 Growth opportunities

E-commerce retailers who analyse and understand market dynamics and customer preferences can identify untapped customer segments and develop innovative products and services to identify new markets, customer segments, and product opportunities.

The iterative DDDM approach enables businesses to refine their strategies and remain competitive in a rapidly changing environment. A major streaming video service uses data analytics to inform aspects of its content creation and market expansion processes.

12.2.3 Strategic inventory management

A multinational retailer uses data to manage its inventory, particularly in preparation for natural disasters. By analysing historical sales data, the organisation discovered that certain products had a significant sales spike before hurricanes. This insight, derived from mining trillions of bytes of sales data, allowed leaders to stock these items heavily in anticipation of storms to meet higher customer demand.

Additionally, the retailer employs predictive analytics to forecast demand for various products based on historical data, weather patterns, and other external factors. So, the company can dynamically adjust its inventory levels to ensure essential items are available when customers need them. Their use of big data and analytics extends to optimising supply chain operations, where real-time data helps the business manage inventory levels across its stores and distribution centres efficiently.

12.2.4 Customer engagement and satisfaction

A global online retailer uses customer data extensively to build targeted marketing campaigns and enhance its recommendation engine. Analysing these data allows the organisation to create personalised shopping experiences and highly targeted marketing campaigns.

In addition to personalising product recommendations, the organisation uses customer data for dynamic pricing strategies. By monitoring competitor prices, market trends, and customer demand in real time, the organisation adjusts its prices to remain competitive and optimise revenue.

12.2.5 Increasing customer retention

A popular online streaming service uses data to personalise recommendations and reduce customer churn. The platform uses vast amounts of customer data, including viewing history, ratings, and even the amount of time spent watching specific content, to tailor its recommendations. This personalisation is driven by sophisticated algorithms that analyse user behaviour to suggest content that aligns with individual preferences.

The organisation employs several strategies to further retain customers and reduce churn. One of the most effective methods is its recommendation algorithm, which continually presents users with content that matches their tastes. This algorithm both recommends what to watch next and also customises the visual presentation of titles to appeal specifically to different viewers. By keeping users engaged with content they are likely to enjoy, they minimise the risk of subscribers cancelling their service.

12.2.6 Proactive business practices

Predictive analytics allow businesses to anticipate trends or challenges and take pre-emptive actions. Financial institutions use advanced machine learning (ML) algorithms to detect and prevent fraud. A proactive approach to combating fraud protects customers from financial loss and enhances trust in the organisation.

Utility companies employ ML and data analytics to predict energy consumption patterns accurately. Analysing large volumes of real-time data involves developing predictive analysis that considers factors such as the time of day, the day of the week, and historical energy loads. Organisations can use the same process in manufacturing and supply chains for real-time, on-demand forecasting.

12.2.7 Better strategic planning

Data insights help formulate realistic strategic plans. A global coffee brand optimises its site selection strategy by using geographic information system (GIS) technology. This technology allows it to analyse local demographics, traffic patterns, and other relevant data; this precise site selection strategy results in improved performance and higher sales for new stores.

12.3 Best Practices for Data-Driven Decision-Making

Every organisation can benefit from taking six steps to integrate data-based decisions throughout the company. By adopting these best practices, strategies derived from data analysis can be implemented and their impact assessed:

Define objectives: This step involves clearly articulating the organisation's goals. After objectives have been defined, the company can make a focused and purposeful effort to meet them.

Identify, prepare, and collect data: In this step, the organisation sets clear objectives, determines data needs, evaluates and prepares data sources, and then systematically gathers and validates the data.

Organise and explore: Here, the data are structured to uncover new patterns, trends, and valuable insights. Cleaning data protects their accuracy and reliability. Visualising the data helps identify patterns, outliers, and trends that are not immediately obvious from raw data.

Perform data analysis: This step transforms raw data into actionable insights by using various techniques and methodologies to uncover patterns, correlations, and trends that inform business strategy. By performing data analysis, organisations can drive strategic decisions and improve overall performance.

Draw conclusions: Here, key data analysis findings are reviewed and placed in the correct business context to form actionable insights and recommendations. Practical conclusions lead to meaningful actions that drive business success.

Implement and evaluate: This final step is crucial for validating insights and measuring outcomes. Action plans are developed based on data-driven recommendations, resources are allocated, and the initiative's progress is continuously monitored. Outcomes are measured against predefined key performance indicators (KPIs), results are analysed, feedback is gathered, and, based on this information, adjustments are made as part of a continuous cycle of monitoring and iterative improvement. This step provides a basis for making informed adjustments to strategies and plans based on empirical evidence and feedback.

12.4 Challenges of Data-Driven Decision-Making

When implementing DDDM, several challenges arise that organisations need to navigate effectively. One significant issue is neglecting data quality. Poor-quality data can lead to inaccurate analyses and misguided decisions, undermining the value of data-driven strategies.

In many organisations, data reside in disparate systems and formats across departments, making it difficult to consolidate and analyse them comprehensively. Implementing data integration solutions is essential to overcoming this hurdle and enabling holistic decision-making.

Data illiteracy is often another critical challenge. Employees may lack the necessary skills to interpret and use data effectively, leading to misinterpretations and suboptimal decisions. Providing ongoing training and fostering a data-literate culture is vital for maximising the benefits of data-driven approaches.

Overreliance on historical data can also be problematic. While past data are valuable, they may not accurately predict future trends, especially in rapidly changing environments. Balancing historical data with real-time analytics and forward-looking indicators is essential for relevant and timely decision-making.

Confirmation bias can skew data analysis. Decision-makers might selectively interpret data to support preconceived notions, leading to biased outcomes. Encouraging objective analysis and critical thinking can help mitigate this bias. Also, poor communication of insights is a similar challenge; even the most accurate data can be ineffective if not communicated clearly and compellingly to stakeholders.

Neglecting data security poses risks. Protecting data from breaches and certifying compliance with data privacy regulations are paramount to maintaining trust and avoiding legal repercussions.

12.5 Types of Data Analysis Used in Data-Driven Decision-Making

Businesses can benefit from understanding the types of business analytics that can inform their decision-making processes.

Descriptive analysis aims to describe and summarise historical data through data aggregation and mining, providing insights into past performance. This type of analysis is used to create monthly sales reports, customer satisfaction surveys, and website traffic analyses.

Diagnostic analysis focuses on determining why certain events occurred. It involves data discovery, mining, and identifying correlations to uncover the root causes of trends or incidents, such as a drop in sales or a surge in customer complaints.

Predictive analysis forecasts future trends or outcomes based on historical data. Businesses can more effectively predict sales and customer behaviour and manage risks by using statistical models, machine learning, and forecasting techniques.

Prescriptive analysis goes a step further by recommending actions based on data. This type combines predictive analytics with optimisation algorithms to suggest the best course of action, aiding in supply chain optimisation, marketing campaign strategies, and resource allocation decisions.

Exploratory analysis is used to discover patterns, relationships or anomalies in data without specific hypotheses. Data visualisation, clustering and dimensionality reduction techniques help identify new market segments, explore customer demographics, and uncover unexpected correlations.

Inferential analysis uses a data sample to make inferences about a population. It employs statistical tests such as regression analysis, hypothesis testing, and confidence intervals and is useful in market research studies, product testing, and consumer preference surveys.

Qualitative analysis focuses on non-numeric data to understand concepts, opinions, or experiences. Methods such as content analysis, thematic analysis, and text mining are employed to analyse customer feedback and social media sentiment and conduct market research interviews.

Quantitative analysis analyses numerical data to quantify variables and uncover patterns using statistical analysis, mathematical modelling, and computational techniques. This type of analysis is essential for financial modelling, operational metrics analysis, and performance measurement.

Real-time analysis involves analysing data as they are generated, providing immediate insights. Using streaming analytics, real-time dashboards, and event processing, businesses can detect fraud, manage inventory in real time, and monitor live customer support.

12.6 Algorithmic Decision-Making

In today's business environment and society, digital data are captured through a variety of devices that have the ability to monitor the minutiae of an individual's everyday life. These data are often processed by algorithms, which support (or drive) decisions (ADM). While the strategic value of these data (and subsequent analysis) for businesses is unquestionable, the implications for individuals and wider society are less clear. ADM refers to the use of computational methods that enable machines to automatically complete tasks and/or make decisions. Algorithms can make decisions because of the vast volumes of electronic data that have become available. ADM involves using computational methods, often AI and ML, to make decisions automatically. These systems analyse data, evaluate options, and select the best course of action based on predefined criteria. ADM is increasingly used in various fields such as healthcare, finance, and transportation, offering potential for efficiency, accuracy, and adaptability.

While algorithms are hardly a recent invention, they are nevertheless increasingly involved in systems used to support decision-making. These systems, known as ADM systems, often rely on the analysis of large amounts of personal data to infer correlations or, more generally, to derive information deemed useful to make decisions.

Human intervention in the decision-making may vary, and humans may even be completely out of the loop in entirely automated systems. In many situations, the impact of the decision on people can be significant, such as access to credit, employment, medical treatment, or judicial sentences. Entrusting ADM to make or influence such decisions raises a variety of ethical, political, legal, or technical issues, where great care must be taken to analyse and address them correctly. If they are neglected, the expected benefits of these systems may be negated by a variety of different risks for individuals (discrimination, unfair practices, loss of autonomy, etc.), the economy (unfair practices, limited access to markets, etc.), and society as a whole (manipulation, threats to democracy, etc.). Decision-making algorithms are increasingly used in areas such as access to information, e-commerce, recommendation systems, employment, healthcare, and justice. Algorithmic decisions can be based on handcrafted systems that use simple scoring mechanisms, the identification of keywords, or natural language extraction. Rules may be articulated directly by programmers or be dynamic and flexible based on ML of contextual data.

These ADM systems often rely on the analysis of large amounts of personal data to infer correlations or, more generally, to derive information deemed useful to make decisions.

Data-driven or algorithmic decision-making is based on collecting and analysing large quantities of data that are then used to make strategic decisions (Newell and Marabelli, 2015). ADM incorporates two main characteristics:

1. Huge amounts of data and information are the foundation for decisions.
2. Reasons, logic, and origins are behind the suggested and used algorithms for decision-making.

12.7 Threats of ADM

ADM could undermine the fundamental principles of equality, privacy, dignity, autonomy, and free will, and it may also pose risks related to health, quality of life, and physical integrity. That ADM can lead to discrimination has been extensively documented in many areas, such as the judicial system, credit scoring, targeted advertising, and employment. Discrimination may result from different types of biases arising from the training data, technical constraints, or societal or individual biases. However, the risk of discrimination related to the use of ADM should be compared with the risk of discrimination without the use of ADS. Humans have their own sources of bias that can affect their decisions, and in some cases, these could be detected or avoided using ADM. The deployment of ADM may also pose a threat to privacy and data protection in various ways. The first is related to the massive collection of personal data required to train the algorithms. Even when no external attack has been carried out, the mere suspicion that one's personal data are being collected and possibly analysed can have a detrimental impact on people. For example, several studies have provided evidence of the chilling effect resulting from the fear of online surveillance. Altogether, large-scale surveillance and scoring could narrow the range of possibilities and choices available to individuals, affecting their capacity for self-development and fulfilment. Scoring also raises the fear that humans are increasingly treated as numbers and reduced to their digital profiles. Reducing the complexity of human personality to a number can be seen as a form of alienation and an offence to human dignity.

Bibliography

Bang, C.G. (2025) *Data-Driven Decision-Making for Business*. 1st Edition. London: Routledge.

Dignum, V. (2019) *Responsible Artificial Intelligence: How to Develop and Use AI in a Responsible Way*. Wiesbaden: Springer Nature.

Newell, S. and Marabelli, M. (2015) 'Strategic Opportunities (and challenges) of Algorithmic Decision-Making: A Call for Action on the Long-Term Societal Effects of Datification', *The Journal of Strategic Information Systems*, 24(1), pp. 3–14. Available at: https://www.sciencedirect.com/science/article/abs/pii/S0963868715000025 (Accessed: 14 March 2025).

Pezotta, G. *et al.* (2024) *Data-Driven Decision Making for Product Service Systems*. Cham: Springer.

Steinhardt, G. (2024) *Data-Driven Decision-making for Product Managers: A Primer to Data Literacy in Product Management (Business Guides on the Go)*. Wiesbaden: Springer.

Zhang, G. and Gao, J.L. (2015) *Multi-Level Decision Making: Models, Methods and Applications*. Heidelberg: Springer.

Chapter 13

AI-Powered Learning Solutions in Higher Education

13.1 Introduction

Artificial intelligence (AI) is increasingly recognised as a pivotal force in the transformation of higher education. As institutions navigate the complexities of globalisation, digitalisation, and a rapidly evolving job market, the integration of AI technologies offers new pathways for addressing persistent challenges in teaching, learning, and administration. AI's capacity to process massive data sets, detect meaningful patterns, and automate tasks provides unprecedented opportunities to enhance educational outcomes, increase institutional efficiency, and personalise learning experiences (Holmes *et al.*, 2019).

The landscape of higher education is undergoing a fundamental shift. Students today expect learning environments that are flexible, adaptive, and technologically enriched. At the same time, educators and administrators are under pressure to do more with less – streamlining operations, supporting diverse learners, and preparing graduates for an AI-driven workforce. AI tools, from intelligent tutoring systems to predictive analytics and virtual assistants, are helping to meet these demands by delivering more tailored and efficient educational services (Luckin *et al.*, 2016).

AI-powered learning solutions are not limited to classroom instruction. They include a spectrum of innovations such as adaptive learning platforms, automated grading systems, chatbots for student support, and systems that identify at-risk students through predictive modelling. These applications are reshaping both pedagogical strategies and institutional structures, enabling a more data-informed and student-centred approach to education (Zawacki-Richter *et al.*, 2019). Importantly, AI can help bridge access gaps by offering tools for students with disabilities and supporting multilingual learning environments.

Yet, the adoption of AI in higher education is not without its complications. Questions about algorithmic bias, data security, and equity remain critical. While AI has the potential to democratise learning, it can also reinforce existing inequalities if not implemented with care. Institutions must therefore adopt ethical frameworks and inclusive practices that guide the responsible use of AI (Roll and Wylie, 2016).

This chapter explores the transformative impact of AI in higher education, examining its role in pedagogy, administration, and student support. It will also consider the ethical and social implications of AI adoption and present a case study illustrating successful implementation in a university setting. In doing so, the chapter aims to provide both a theoretical and practical understanding of how AI can support innovation and equity in higher education.

The conceptual diagram shown in Figure 13.1 illustrates the interconnected components of an AI-powered higher education system, including adaptive learning, predictive analytics, virtual assistants, and accessibility tools.

13.2 The Role of AI in Modern Higher Education

AI is not merely a technological innovation; it is a transformative force reshaping higher education in profound ways. At the core of this transformation is AI's ability to process vast amounts of data, detect patterns, and automate complex processes with remarkable precision and speed. These capabilities enable institutions to tackle longstanding challenges, enhance operational efficiency, and elevate the quality of learning experiences to levels previously thought unattainable. As higher education continues to evolve, AI is not merely complementing traditional methods but is also driving entirely new approaches to teaching, learning, and institutional management.

This transformative impact is particularly evident in the customisation of learning experiences. AI systems analyse millions of data points from diverse student populations, identifying trends that allow educators to tailor their methods to individual needs. For example, AI can recommend

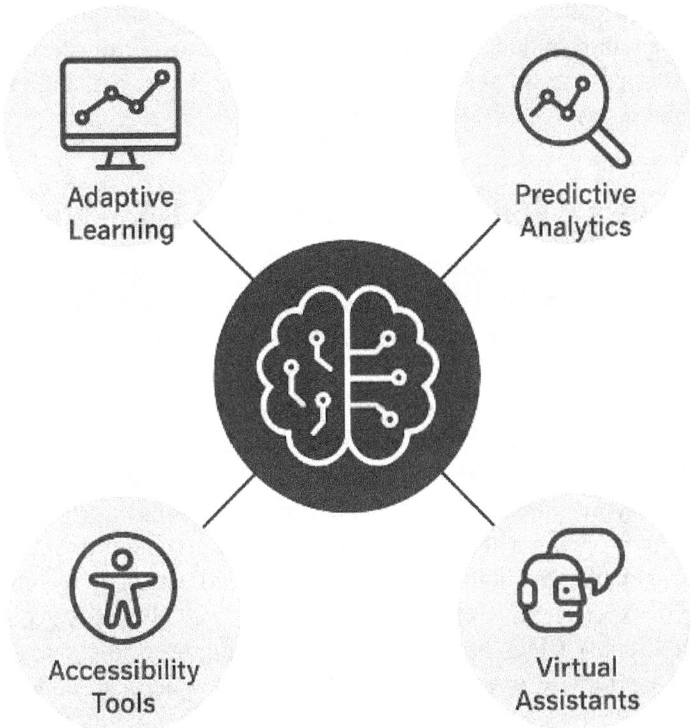

Figure 13.1. Components of an AI-powered learning ecosystem.

specific learning pathways for students struggling in certain subjects, ensuring they receive the necessary support to succeed. Furthermore, institutions leverage AI to design predictive models that inform strategic planning, enabling them to address potential challenges proactively, such as enrolment fluctuations or resource allocation inefficiencies.

In addition to its impact on learning, AI is playing a pivotal role in fostering global collaboration within the academic community. By breaking down language barriers with real-time translation tools and enabling seamless virtual interactions, AI facilitates a more interconnected world of higher education. This increased connectivity empowers institutions to share resources, exchange innovative teaching practices, and collectively address global challenges such as sustainability and diversity in education.

This section explores how AI is becoming an integral part of modern higher education, not only through its immediate applications but also by shaping the broader landscape of learning and institutional operations. From enhancing student experiences to redefining administrative processes, AI is paving the way for a more dynamic, inclusive, and efficient educational ecosystem.

13.3 AI as a Catalyst for Educational Innovation

AI-powered tools and systems are driving innovation across higher education, redefining both teaching and learning. One of the most significant impacts of AI is its ability to personalise education. Traditional education models often rely on a one-size-fits-all approach, which can fail to meet the diverse needs of students. AI, however, enables adaptive learning systems that tailor instructional content and delivery to individual learners. By analysing real-time data, these systems can adjust difficulty levels, recommend supplementary resources, and provide targeted feedback, ensuring that each student progresses at their own pace (Zawacki-Richter *et al.*, 2019).

Moreover, AI facilitates active and engaged learning by enabling immersive experiences. Technologies such as virtual reality (VR) and augmented reality (AR), powered by AI, are being used to create simulations and interactive environments. For instance, medical students can practise surgical procedures in virtual operating rooms, while engineering students can experiment with structural designs in virtual laboratories. These applications enhance experiential learning, making complex concepts more accessible and engaging for students.

13.4 Operational Efficiency and Institutional Transformation

Beyond the classroom, AI is revolutionising administrative processes within higher education institutions. From admissions to enrolment and scheduling, AI systems streamline operations that were once time-intensive and prone to errors. For example, natural language processing (NLP)-based chatbots are now being deployed to handle routine enquiries, provide campus information, and assist with application processes, reducing the workload on administrative staff. These tools enhance responsiveness and ensure that students receive timely support (Holmes *et al.*, 2019).

Predictive analytics, another key application of AI, is being used to identify trends and anticipate challenges. By analysing historical and real-time data, institutions can predict enrolment patterns, optimise resource allocation, and develop strategies to improve student retention. This data-driven approach enables universities to make informed decisions that enhance both efficiency and student success. For instance, AI algorithms can detect early signs of academic difficulties, allowing advisors to intervene with tailored support programmes.

13.5 Enhancing Accessibility and Equity

One of the most promising aspects of AI in higher education is its potential to make learning more inclusive. AI-powered tools can break down barriers for students with disabilities, providing features such as speech-to-text transcription, text-to-speech conversion, and real-time language translation. These tools ensure that all students, regardless of their physical or linguistic limitations, have equal access to educational opportunities (Luckin *et al.*, 2016). Moreover, adaptive technologies can address disparities in learning resources, providing personalised support for students from underrepresented or disadvantaged backgrounds.

However, achieving equity through AI also requires addressing the digital divide. Not all institutions or students have equal access to advanced technologies, and this gap can exacerbate existing inequalities. To realise the full potential of AI in creating equitable learning environments, policymakers and educators must invest in infrastructure, training, and digital literacy programmes.

13.6 Key Applications of AI in Higher Education

AI has introduced a wide range of applications that are reshaping both the academic and administrative dimensions of higher education. These applications are not only enhancing learning outcomes but also improving institutional effectiveness and student engagement. In this section, we explore several core areas where AI is having the most significant impact.

13.6.1 Personalised and adaptive learning

Adaptive learning technologies, powered by AI, are transforming the one-size-fits-all approach to instruction. These platforms assess individual learner performance in real time and adjust content accordingly.

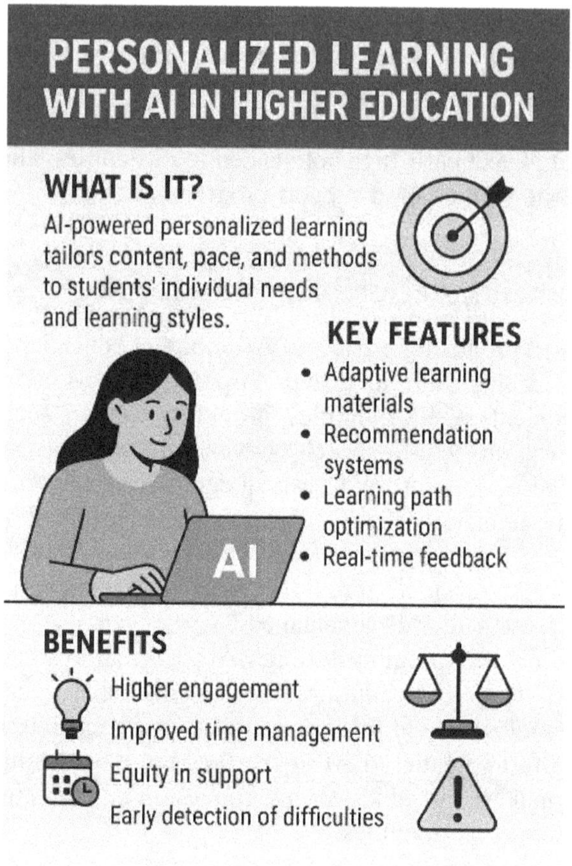

Figure 13.2. Personalised learning with AI in higher education.

By continuously analysing data, such systems can recommend additional resources, alter the sequence of instructional material, or even change the modality of delivery to match a student's preferences or learning style (Figure 13.2). This leads to increased learner satisfaction and improved academic achievement (Chen *et al.*, 2020).

13.6.2 AI-supported assessment and feedback

AI tools are being used to automate the grading of assignments, especially in large-enrolment courses. These systems not only provide instant feedback to students but also help instructors identify learning gaps and misconceptions. For example, AI can analyse patterns in student responses to highlight areas where the cohort as a whole struggles, thus enabling instructors to adapt their teaching strategies. NLP allows AI to assess short written responses, offering nuanced feedback in formative assessments (Roll and Wylie, 2016).

13.6.3 Virtual assistants and student support

AI-driven virtual assistants, such as chatbots, are being widely implemented to offer 24/7 support for students. These tools handle a broad spectrum of queries, from logistical information such as deadlines and timetables to academic support such as course recommendations or the clarification of concepts. The availability of such immediate support has been shown to reduce administrative bottlenecks and increase student satisfaction (Holmes *et al.*, 2019).

13.6.4 Predictive analytics and learning analytics

AI systems equipped with predictive analytics can identify students who are at risk of academic failure or dropout by analysing attendance records, assignment submissions, and other behavioural data. Early interventions can then be initiated, often through personalised outreach or targeted support programmes. Learning analytics dashboards provide educators and advisors with visual summaries of student progress and engagement, making it easier to take timely and informed actions (Zawacki-Richter *et al.*, 2019).

13.7 Ethical Considerations and Equity

While AI presents numerous benefits, it also raises complex ethical issues that must be carefully addressed. These concerns include algorithmic bias, data privacy, and equitable access to technology.

13.7.1 Algorithmic bias and fairness

AI systems are only as unbiased as the data on which they are trained. When these data sets contain historical inequalities or reflect systemic biases, the resulting algorithms can perpetuate or even amplify these

Table 13.1. Key applications and functions of AI in higher education.

AI Application	Function and Benefit
Adaptive Learning Systems	Personalise content delivery based on learner performance.
Automated Assessment	Reduce grading time and provide immediate, data-informed feedback.
Virtual Assistants	Offer 24/7 support for administrative and academic questions.
Predictive Analytics	Identify at-risk students and support early intervention strategies.
Learning Analytics	Visualise student engagement and academic trends.

disparities. For example, predictive models might unfairly flag students from underrepresented backgrounds as being "at risk" due to correlations in historical data (Table 13.1). Institutions must rigorously audit their AI tools and ensure transparency and accountability in their design and use (Eubanks, 2018).

13.7.2 Data privacy and security

AI tools often rely on extensive personal data, including academic records, behavioural data, and sometimes biometric information. Ensuring the security and ethical use of these data is paramount. Institutions must implement robust data governance policies that comply with regional and international standards, such as GDPR in Europe or FERPA in the U.S. Transparent consent mechanisms should be put in place so that students understand what data is collected and how it is used (Slade and Prinsloo, 2013).

13.7.3 Equity and digital inclusion

The implementation of AI could either mitigate or exacerbate existing inequalities in higher education. Students from lower socioeconomic backgrounds may lack access to the necessary digital infrastructure or digital literacy skills to benefit fully from AI-powered tools. Ensuring equitable access to technology and designing inclusive AI systems are critical steps towards reducing these disparities. This includes offering low-bandwidth solutions, supporting multilingual platforms, and providing assistive technologies for students with disabilities (Williamson and Piattoeva, 2021).

13.8 Future Trends and Impacts

As the integration of AI into higher education continues to deepen, several emerging trends are poised to shape the sector in the coming decade. These trends reflect not only technological advances but also evolving pedagogical philosophies and societal needs.

13.8.1 Generative AI and content creation

The rise of generative AI tools, such as large language models, opens new possibilities for content development and curriculum design. Educators can use these tools to generate quiz questions, summarise articles, and even draft syllabi. Students, in turn, can use generative AI to enhance their learning by experimenting with interactive explanations, simulated dialogues, or customised study guides. However, institutions must develop academic integrity policies that address the ethical use of such tools in academic work.

13.8.2 AI and lifelong learning

As the nature of work continues to evolve, higher education institutions are increasingly expected to support lifelong learning. AI can facilitate this by offering flexible, modular, and personalised learning pathways that adapt to learners' evolving needs over time. Micro-credentialing and just-in-time learning solutions powered by AI are helping professionals upskill and reskill throughout their careers (Jaschik, 2022).

13.8.3 Institutional transformation and strategy

AI is prompting higher education institutions to rethink their organisational strategies. From investment in digital infrastructure to the development of AI literacy among faculty and students, institutions are being reshaped at all levels. Strategic planning now includes considerations for ethical AI governance, data management, and cross-disciplinary collaboration between technologists and educators. These changes suggest that the most successful institutions in the future will be those that align their missions with the ethical and innovative use of AI.

13.9 Case Study: IU International University and the Implementation of Synthea

IU International University, Germany's largest university, has emerged as a frontrunner in the integration of AI into higher education. With a student body exceeding 100,000 learners, the institution has prioritised scalable, innovative, and data-driven strategies to support personalised learning at scale. Central to this transformation is IU's proprietary AI-driven learning platform, Synthea, which has redefined how educational content is delivered, assessed, and optimised.

13.9.1 Context and rationale

IU's rapid growth in online and blended learning models presented both opportunities and challenges. The university needed a robust system capable of delivering personalised instruction to tens of thousands of students while maintaining pedagogical quality and academic rigour. Synthea was developed to meet this need a machine learning-based platform designed to enhance learner engagement, streamline content adaptation, and facilitate real-time performance analytics.

13.9.2 Implementation and functionality of Synthea

Synthea operates at the intersection of adaptive learning and intelligent tutoring. It analyses vast data sets from students' learning behaviours – such as quiz performance, engagement rates, and navigation patterns – to tailor content pathways. This ensures that learners receive material aligned with their current level of mastery and learning pace. The system continuously adapts based on performance, offering dynamic recommendations and real-time feedback.

A distinctive feature of Synthea is its modularity. The platform breaks down academic content into micro-learning units, allowing students to progress through clearly defined, manageable chunks of information. Each module is accompanied by formative assessments, immediate AI-generated feedback, and tailored content reinforcement, all of which are continuously refined through learning analytics.

Furthermore, Synthea supports faculty by generating insights into individual and group performance trends. Instructors receive dashboards highlighting students who are struggling or excelling, enabling data-informed interventions. The platform also offers predictive analytics to forecast course completion risks, empowering academic advisors to take proactive measures.

13.9.3 Impact and outcomes

Since its implementation, Synthea has significantly improved both learner satisfaction and academic performance at IU. Internal evaluations show increased student engagement, higher course completion rates, and improved learner confidence in self-paced environments. The platform has also enabled IU to scale without sacrificing quality, offering personalised education to a diverse, geographically dispersed student body.

From an institutional perspective, Synthea has streamlined content development and curriculum management. Faculty members report reduced administrative burdens, allowing more time for mentoring and research. Moreover, the data collected through Synthea inform strategic planning and curriculum improvement, creating a feedback loop that aligns institutional goals with student success metrics.

13.9.4 Lessons learned and broader implications

IU's experience with Synthea illustrates the transformative potential of AI in higher education when strategically implemented. Key success factors included strong institutional leadership, investment in in-house technological development, and a focus on pedagogical alignment. However, the case also underscores the importance of transparency, data ethics, and ongoing evaluation in AI deployments.

As other institutions look to replicate similar innovations, IU's model offers a blueprint for scaling personalised learning through AI. The successful deployment of Synthea demonstrates how universities can harness data-driven approaches to enhance educational quality, accessibility, and efficiency on a national and global scale.

13.10 Conclusion and Recommendations

This chapter has examined the transformative role of AI in higher education, highlighting its applications, challenges, and future potential. AI is enabling a shift from standardised to personalised learning by adapting instructional content in real time to meet individual student needs (Chen *et al.*, 2020). Through tools such as predictive analytics, AI also allows institutions to identify and support at-risk learners early, improving retention and completion rates (Zawacki-Richter *et al*., 2019).

Administrative functions have seen considerable benefits as well. AI technologies streamline complex processes such as admissions, scheduling, and grading, thus freeing up time for educators and staff to focus on more strategic and student-centred activities (Holmes, Bialik and Fadel, 2019). Meanwhile, virtual assistants and AI-supported chatbots provide scalable, 24/7 support, significantly enhancing the student experience.

Ethical issues remain critical, including the risks of bias embedded in algorithms, the need for robust data privacy policies, and concerns about equitable access to AI-enhanced education (Eubanks, 2018; Slade and Prinsloo, 2013). The case of IU International University shows how these challenges can be addressed through careful platform design, institutional commitment, and a strong focus on educational equity.

AI's potential in supporting lifelong learning and global collaboration further underscores its importance as a strategic asset in higher education. However, the success of these initiatives depends on intentional implementation, ethical oversight, and inclusive access to digital infrastructure (Williamson and Piattoeva, 2021).

The integration of AI into higher education is not a distant possibility; it is a current and rapidly evolving reality. As this chapter has illustrated, AI offers transformative potential across multiple domains, from personalised learning and predictive analytics to administrative automation and enhanced accessibility. Institutions such as IU International University exemplify how AI can be harnessed at scale to improve learning outcomes, operational efficiency and student satisfaction.

However, with opportunity comes responsibility. The deployment of AI in educational contexts must be guided by ethical principles that ensure fairness, transparency, and inclusivity. Institutions must be vigilant against algorithmic bias, protect student data privacy, and bridge digital divides that could otherwise exacerbate educational inequalities. The successful implementation of AI thus requires not only technological investment but also strong leadership, institutional vision, and continuous stakeholder engagement.

As AI technologies continue to advance, several key recommendations emerge for higher education institutions:

1. **Develop institutional AI strategies:** Institutions should formulate clear strategies that align AI adoption with their mission and pedagogical goals.

2. **Prioritise ethical governance:** Establish ethical frameworks to guide the design, deployment, and evaluation of AI systems.
3. **Invest in faculty and staff training:** Equip educators and administrators with the digital and data literacy needed to use AI effectively and responsibly.
4. **Foster interdisciplinary collaboration:** Encourage collaboration between computer scientists, educators, sociologists, and ethicists to develop inclusive and effective AI tools.
5. **Evaluate and iterate continuously:** Use learning analytics and user feedback to refine AI applications and ensure their relevance and effectiveness.

Ultimately, AI should not be seen as a replacement for human educators but as a powerful ally in the pursuit of accessible, high-quality, and equitable education. By leveraging AI thoughtfully and ethically, higher education institutions can respond to contemporary challenges with agility and vision – preparing learners not only for today's knowledge economy but also for the complexities of a digitally mediated future.

Bibliography

Chen, G., Fan, X. and Lin, T. (2020) 'Adaptive Learning Platforms in Higher Education: Enhancing Retention Through Personalized Learning Pathways', *Journal of Educational Technology Research*, 28(3), pp. 125–140.

Eubanks, V. (2018) *Automating Inequality: How High-Tech Tools Profile, Police, and Punish the Poor*. New York: St. Martin's Press.

Holmes, W., Bialik, M. and Fadel, C. (2019) *Artificial Intelligence in Education: Promises and Implications for Teaching and Learning*. Boston: Center for Curriculum Redesign.

IU International University (2023) *Synthea: AI-Powered Learning Innovation at IU*. Internal White Paper.

Jaschik, S. (2022) 'The AI-driven Campus: Emerging Trends and Strategic Directions', *Chronicle of Higher Education*, 68(14), pp. 21–28.

Luckin, R. *et al*. (2016) *Intelligence Unleashed: An Argument for AI in Education*. London: Pearson Education.

Roll, I. and Wylie, R. (2016) 'Evolution and Revolution in Artificial Intelligence in Education', *International Journal of Artificial Intelligence in Education*, 26(2), pp. 582–599.

Slade, S. and Prinsloo, P. (2013) 'Learning Analytics: Ethical Issues and Dilemmas', *American Behavioral Scientist*, 57(10), pp. 1510–1529.

Williamson, B. and Piattoeva, N. (2021) 'Education Governance and Datafication: Reimagining Learning, Teaching and Research', *European Educational Research Journal*, 20(4), pp. 496–512.

Zawacki-Richter, O. *et al.* (2019) 'Systematic Review of Research on Artificial Intelligence Applications in Higher Education', *International Journal of Educational Technology in Higher Education*, 16(1), 39.

Chapter 14

The Future of AI: Cutting-Edge Trends in AI

The development of full artificial intelligence could spell the end of the human race. It would take off on its own, and re-design itself at an ever-increasing rate. Humans, who are limited by slow biological evolution, couldn't compete, and would be superseded.

— Stephen Hawking

14.1 AI Will Change Organisations, Systems, and People

When it comes to AI, such shocking monstrosities are uttered so routinely that they almost sound boring. For example, "machines will soon be able to think as well as humans". Anyone who informs themselves about AI and speaks to people in the field reads and hears this all the time. Often, in the abbreviated form, "AGI is coming, general artificial intelligence is not far off". Hardly anyone in the industry seems to doubt this; the only question remaining is how long it will take. Outside the tech bubble, things feel different. Yes, some people have vague fears. But my impression is that many people, especially in Germany, aren't particularly interested in whether AGI is actually imminent, and if they do think about it, they assume the announcements are exaggerated.

Since its introduction, generative AI (Gen AI) has evolved. We have already seen developers such as OpenAI and Meta move away from large models to improve AI models to do the same or more with less. Prompt engineering is evolving as models like ChatGPT become increasingly intelligent and better able to understand the nuances of human language. As LLMs are trained on more specific information, they can provide in-depth expertise to specialised industries and become highly available agents to help complete tasks.

AI is not a flash in the pan. It's not a phase. Over 60 countries have developed national AI strategies to harness the benefits of AI while mitigating its risks. This means significant investment in research and development, reviewing and adapting relevant policy standards and regulatory frameworks, and ensuring that the technology does not decimate fair labour markets and international collaboration.

Human–machine communication is becoming increasingly easier, enabling AI users to do more and better work. Through continuous research and optimisation, AI is expected to contribute US$4.4 trillion to the global economy. By 2034, AI will be an integral part of many aspects of our personal and business lives. Generative AI models such as GPT-4 have shown enormous potential in the short time they have been publicly available, but their limitations are also well known. As a result, the future of AI will be defined by a shift towards open-source large-scale models for experimentation and the development of smaller, more efficient AI models to increase usability and enable lower costs (Mucci and Gather, 2025). Initiatives such as Llama 3.1, an open-source AI model with 400 billion parameters, and Mistral Large 2, released for research purposes, illustrate the trend towards encouraging community collaboration in AI projects while preserving commercial rights. Growing interest in smaller models has led to the development of models such as the 11-billion-parameter MGPT 4o mini, which is fast and cost-effective. It won't be long before there's a model suitable for embedding in devices like smartphones, especially as costs decrease (Mucci and Gather, 2025).

This movement reflects the transition from exclusively large, closed models to more accessible and versatile AI solutions. While smaller models offer affordability and efficiency, there remains public demand for more powerful AI systems, suggesting that there will likely be a balanced approach to AI development that seeks to prioritise both scalability and accessibility. These new models offer greater precision with fewer resources, making them ideal for companies that need to create customised content or require complex problem-solving capabilities. AI has influenced the development of several core technologies. AI plays a critical role in the advancement of computer vision by enabling more accurate image and video analysis, which is essential for technologies such as autonomous vehicles and medical diagnostics. In natural language processing (NLP), AI improves the ability of machines to understand and generate human language, improves communication interfaces, and enables more sophisticated translation and sentiment analysis tools.

AI enhances predictive and big data analytics by processing and interpreting massive amounts of data to predict trends and inform decisions. In robotics, the development of autonomous and more adaptable machines simplifies tasks such as assembly, exploration, and service delivery. Furthermore, AI-powered innovations in the Internet of Things (IoT) improve the connectivity and intelligence of devices, leading to smarter homes, cities, and industrial systems.

14.2 AI as a Lever for New IT Strategies

Against the backdrop of the AI revolution, the focus of the IT function is fundamentally changing. While previously a key driver of digital transformation, the department is now broadening its focus to include AI transformation within the company. Through innovative, productive use of AI, CIOs are pioneering work for the entire organisation and inspiring other leaders to pursue transformative initiatives.

What impact does this have on the IT department itself? GenAI is now increasingly being used for efficient development, and engineering is increasingly moving away from previous manual tasks in the software development cycle. The code itself is increasingly being generated by AI assistants, and human review is carried out for quality assurance (a human-in-the-loop process). IT staff are now increasingly devoting themselves to architecture, orchestration, and code review. Since the corresponding AI skills continue to be scarce within their own teams and on the job market, companies are turning to upskilling approaches, which in turn utilise AI-based training methods. AI tools also help with cost optimisation, for example, for cloud services (Cloud Financial Operations and FinOps). AI-based infrastructure automation also enables significantly more efficient and flexible operations. However, new cyber risks must also be addressed through GenAI. Overall, thanks to AI, IT workers will have significantly more capacity for more demanding tasks with a higher value contribution, whereby they will increasingly have to contribute their own business understanding. Productivity in this department will increase significantly. The transformation of the IT function could go so far that, with its advanced AI capabilities, it could enable an internal, highly efficient "IT-as-a-service" approach for the entire organisation. In an outlook, several reports and forecasts outline the far-reaching implications of this blurring of boundaries within the technology function (Helmold, 2024).

14.3 Generative AI Will Change the Jobs of Tomorrow

GenAI has already fundamentally altered the way many jobs are performed. Thanks to its ability to generate content that profoundly supports human expertise and skills, it is transforming many aspects of our work. With its potential to support a wide range of output formats – from audio and video to text and code and even 3D models – GenAI is transforming a multitude of tasks. Whether writing reports, designing website graphics, creating personalised marketing strategies, or developing learning programmes, the use cases of GenAI in various industries are limitless and illustrate the enormous potential of this technology.

With the forecast that 40 percent of enterprise applications will soon have embedded conversational AI and that nearly 15 percent of new applications will be autonomously generated by AI by 2027, there is no question that a significant shift in the future of work is imminent. This makes it clear that our relationship with technology and the design of work ecosystems must be redefined.

We are therefore faced with a crucial question: How must work be rethought in order to make the most of this transformation?

Both GenAI and general trends such as the shift towards skill-based organisations are influencing the way we work. Against this backdrop, we must rethink our traditional understanding of work. Today, work encompasses not only human skills but also the contributions of AI. Jobs are losing their traditional meaning as fixed roles and are becoming dynamic functions that constantly evolve depending on the demands of the work environment.

In this dynamic work environment, the performance of certain tasks is becoming increasingly complex. It requires the adaptation and development of skills and methods to achieve a specific result. GenAI offers the potential to automate these tasks. This can free up employees' skills to focus on new tasks or make those tasks easier for employees, creating more time for individuals.

While some tasks are better performed by humans and others by machines, there are also tasks that require a combination of both. According to Deloitte's Skill-Based Organization Survey, companies that invest in skill-based structures achieve 63 percent better results than those with job-based practices.

In this transformation process, forward-thinking companies are increasingly abandoning rigid job titles and instead relying on flexible, skills-based teams. This enables employees to perform different tasks as needed while preparing the company for future changes. This strategic realignment of the workforce, which emphasises dynamic skills and goal-oriented results, paves the way for an organisation that ensures consistency and traction in an ever-changing world of work. This raises the question: How can we maximise the potential and contributions of each individual employee in the context of this transformation?

A human-centred approach is essential to meeting this challenge. This approach puts people at the centre, leveraging both the strengths of generative AI and unique human capabilities. Despite technological advancements, traits such as emotional intelligence, critical thinking, leadership skills, and problem-solving abilities are difficult to replicate in machines. These human skills are essential for building an adaptable and growth-oriented workforce that can thrive in an ever-changing and disruptive world of work. Companies must develop innovative approaches to future-proof their workforce in an AI-driven world of work and meet the evolving demands.

14.4 AI Will Change the Concept of Leadership

Based on the need to put people at the centre, a realignment of leadership roles is essential. Leaders face the challenge of steering their organisations with a clear vision and strategic understanding, without possessing in-depth AI expertise. Their primary responsibility lies in creating a supportive framework for AI-induced change (Hohoff, 2024). This requires leaders to recalibrate their approach and have the courage to venture into uncharted territory. The transformation rests on three fundamental pillars:

1. This begins with adopting an explorer's mindset to understand and experiment with GenAI.
2. This is followed by co-creation to redefine roles and harness the collective potential of humans and machines.
3. Finally, prioritising human performance is critical to shaping an AI-driven world that is both efficient and human-centric.

Given the rapid development of GenAI, leaders must move beyond purely reactive modes and become active designers of the AI-oriented environment. Combining the above-mentioned pillars forms the foundation for thriving in a technology-driven world. Leaders have a responsibility to secure this development and the resulting growth for themselves, their teams, and their organisations (Staff and Groß, 2024). GenAI opens up the possibility of deeper interaction between people and technology and the opportunity to redesign our working world. How we use this technology and integrate it into our processes is in our hands. The forthcoming phases of change should not be viewed as obstacles but as challenges that we must address. Successfully overcoming them can serve as a catalyst for growth and innovation.

The demand for GenAI tools in companies is growing, leading to a world in which traditional work processes are being disrupted and new ways of working are emerging. In this transformative environment, increasingly shaped by AI, a new compass of skills for leaders is crucial (Staff and Groß, 2024). A holistic view of leadership skills in relation to GenAI can be obtained by illuminating multiple dimensions on an affective, data-centric, cognitive, and ethical level.

14.5 Robots Will Revolutionise Our Way of Living

Robotics in everyday life is rapidly changing our lives: Intelligent systems are increasingly supporting us with everyday tasks, saving time, providing freedom, and opening up new possibilities in household, healthcare, and mobility. Smart assistants create efficient solutions and promote independence today more than ever (OECD, 2020).

The key points are summarised in Table 14.1.

Table 14.1. AI trends.

Daily Human Support	Household robots take over routine tasks, from cleaning to gardening.
Healthcare Service	Care robots relieve staff and support elderly people in their daily lives.
Human–Robot Interactions	Humanoid machines communicate at eye level and promote social interaction.
Learning and Improving	Artificial intelligence makes robots capable of learning and flexible deployment.
Smart Technologies Further Develop and Enhance Robots	Smart technologies enable more inclusive lifestyles and greater autonomy.

14.5.1 Household robots: Intelligent helpers in everyday life

Robots are no longer merely a gadget – they have become indispensable in many homes. Robot vacuums and floor mops, in particular, are now standard equipment in modern households. But they can now do much more than simply clean floors. AI-controlled household helpers drive themselves to their charging stations, recognise carpets, communicate via an app or voice, and even learn from mistakes. This allows them to significantly improve their behaviour over time.

New additions include robots that clean windows, mow lawns, sort washing machines, or even fold clothes. In some households, they assist with childcare or pet feeding. The variety of their applications continues to grow rapidly. Some models also serve as security monitors, detect unusual noises, and notify the police or emergency services if necessary.

14.5.2 Robotics in healthcare management: Assistance that relieves the burden

Demographic changes present nursing facilities with significant challenges. To ensure the care of older people, care robots are taking on an increasing number of tasks. These include the physically demanding transfer of patients, the distribution of medications, and simple delivery services. Emotional support provided by social robots such as Paro – a seal-like robot that recognises touch and reacts to human proximity – is particularly popular. Sensor-equipped wheelchairs and automatic assistance systems help maintain mobility. These devices detect obstacles, adapt driving behaviour, and connect to emergency call systems (Bartneck *et al.*, 2024). Combined with speech and image recognition, these technical assistants are capable of mastering complex situations and effectively relieving the burden on nursing staff.

14.5.3 Humanoid robots: Between technology and empathy

Unlike traditional machines, humanoid robots are not only meant to help but will also act in a proactive way. Robots are also meant to understand, react, and build trust. These robots resemble humans in appearance and movement and thus act more naturally in interactive situations (Bartneck *et al.*, 2024). Thanks to electronic skin, modern prototypes register touch, heat, or pressure. They use these data to respond more empathetically to people. They are being developed specifically for tasks where interpersonal proximity is important, such as in elderly care or as companions for people with mental health issues.

With their human-like gestures and facial expressions, they appear less threatening and are easier to integrate into everyday environments. Voice assistance, learning capabilities, and programmable behaviour make them socially interactive helpers with a growing influence on daily life.

14.5.4 Versatile use through technological advancements

Technological advancements are advancing rapidly. Robots today sense their surroundings using high-resolution sensors and analyse movements, speech, or even facial expressions. This creates natural human–machine communication that goes beyond simple commands. With autonomous self-diagnosis, robots report problems independently or perform maintenance tasks independently.

Equally important are advances in energy concepts and mobility. New battery technologies enable longer operation, while AI ensures faster adaptation to new environments and tasks. This interplay makes modern robots reliable partners in everyday scenarios.

Especially in networked environments – such as smart homes – robots achieve peak performance when they can communicate with other

Table 14.2. Types of algorithms.

Predictive Algorithms	These forecast future events based on historical data, such as predicting crime hotspots or identifying which social services are needed in certain areas.
Prescriptive Algorithms	These recommend specific actions based on data inputs, such as suggesting policy changes or optimising public transportation routes.
Descriptive Algorithms	These provide insights into current or past states, such as summarising the impact of recent policies or reporting on public health trends.

systems. This creates complete smart living solutions that take convenience and security to a new level.

14.6 Algorithms Will Influence Decisions

In the future of work, algorithms will provide many beneficial applications in several industries and areas. While algorithm-enabled systems and devices may reduce sources of human error and enhance worker safety and health, algorithms may also introduce new sources of risks to worker well-being.

Algorithms offer the promise of data-driven decision-making, which can enhance objectivity and consistency. By relying on empirical evidence rather than subjective judgement, algorithmic systems aim to reduce bias and ensure that decisions are grounded in quantitative analysis. For instance, in resource allocation, algorithms can analyse patterns and needs more systematically than human administrators, potentially leading to more equitable distribution of resources (Gamito and Eber, 2020).

Table 14.2 shows the types of algorithms which will be used in the future for decision-making.

14.7 Use of Generative AI Will Further Increase in Society and Organisations

GenAI trends in 2025 include increased adoption across industries, a focus on personalisation and multimodality, and a growing emphasis on ethics and sustainability. Businesses are prioritising GenAI tools, investing in training and hiring AI professionals, and exploring partnerships to accelerate deployment (Verma *et al.*, 2024). Additionally, the technology is being used in diverse areas such as creative content generation, customer experience, and even in fields like healthcare and finance. GenAI is not merely a trend but a transformative force reshaping industries and creating new opportunities (Dhivya *et al.*, 2021). As the technology matures, it will be crucial to address ethical considerations, prioritise talent development, and embrace the potential for AI to drive innovation and productivity across various sectors.

The major trends are summarised as follows:

1. Increased adoption and budget prioritisation: GenAI is rapidly becoming a top priority for businesses, with a significant portion of organisations prioritising it in their IT budgets. This is reflected in the increasing number of chief AI officers (CAIOs) being appointed, signifying a commitment to integrating AI into organisational structures.
2. Focus on personalisation and multimodality: GenAI is being used to create personalised experiences for users across various platforms, from product recommendations to adaptive user interfaces. Multimodality, which allows AI to process and generate content across different formats (text, images, and audio), is also gaining prominence.
3. AI-enhanced creative content generation: GenAI is transforming the creative industries by enabling the generation of hyperrealistic visuals, dynamic videos, and original content. This includes applications in fashion design, game development, and even art and design.
4. AI agents and autonomous systems: AI agents, which are autonomous systems capable of performing specific tasks, are emerging as a key trend. These agents can assist with various tasks, from making travel reservations to managing banking transactions.
5. Ethics, sustainability, and responsibility: As GenAI becomes more pervasive, there's a growing emphasis on ethical considerations, including data privacy, bias mitigation, and responsible AI development. Sustainability is also becoming a focus, with companies exploring how AI can help address environmental challenges.
6. Integration and partnership: Companies are increasingly using a hybrid approach, combining customised off-the-shelf models with support from third-party vendors to streamline the deployment process.

7. Upskilling and talent development: The demand for AI-skilled professionals is growing, leading to increased investment in internal training programmes and external hiring.
8. AI in specific industries: GenAI is being adopted across various industries, including healthcare (drug discovery and personalised medicine), finance (risk management and report generation), and education (individualised learning materials).
9. Growth and market expansion: The GenAI market is experiencing significant growth, with forecasts predicting substantial increases in market size and value.

14.8 Low-Code and No-Code Will Democratise Software Development

In the future, software development will experience a turning point thanks to low-code and no-code tools. These technologies enable even non-technical users to create software and apps, increasing productivity and transforming the development of digital solutions. Skills shortages and overburdened IT departments are driving demand for such tools. AI-powered coding tools such as GitHub Copilot and Spark support developers with complex tasks and enable the creation of micro-apps without programming knowledge. Low-code platforms promote the concept of "citizen development", which enables employees without an IT background to develop solutions independently. Forecasts suggest that over 70 percent of application development could be based on low-code by 2025. Despite the benefits, there are concerns about a potential over-dependence on these tools and the neglect of basic programming skills. Companies must ensure that these tools do not replace professional experience and training. Balancing innovation and a sense of tradition remains a key challenge.

14.9 AI Avatars Will Shape New Creative and Ethical Standards

Voice cloning, generative video AI, and multimodality are revolutionising the creative industry by enabling the creation of realistic digital avatars. These technologies are used in areas such as marketing, entertainment, and education, but they raise ethical and legal questions, particularly regarding deepfakes and copyright. While companies like Zoom and Microsoft are working on AI-powered avatars that can be used in meetings, projects such as the AI-generated Jesus avatar in Lucerne demonstrate the true versatility of digital representatives (Helmold, 2023, 24). Studios in the music and film industries are also already utilising AI tools for production and editing, leading to debates about their influence on human artists. One thing is clear: The technology opens up new possibilities but also poses risks for disinformation and misuse.

14.10 Automated Learning Platforms Will Democratise Education

The increasing integration of AI into the educational landscape promises a fundamental transformation of learning. Highly personalised learning platforms and automated content creation are intended to democratise education and make it accessible and tailored to individual needs. In developing regions, such technologies could revolutionise access to education. Despite their potential, there are challenges such as "toolification" and the skills paradox that must be overcome. Critical thinking and media literacy are crucial for using AI effectively. However, the circumstances under which AI stifles creativity or encourages cheating require further investigation. One thing is clear: Educational institutions must rethink traditional learning concepts and foster a deep understanding of the technologies and ethical issues to harness the positive potential of AI.

14.11 LAMs and CUAs Will Take Control of Your Desktop

Large action models (LAMs) and computer-using agents (CUAs) are further developments of large language models (LLMs) that go beyond pure text generation and can utilise tools (Dhivya *et al.*, 2021). These interactive systems could soon independently control many of our typical PC tasks by interpreting queries, analysing various data types, and executing targeted actions (Helmold, 2024). How does this work? The systems combine neural networks with symbolic reasoning to make logical decisions and automate complex tasks.

14.12 Increasing Human–Machine Symbiosis

The interconnectedness of humans, devices, and machines is omnipresent in everyday life. Devices and machines are informing and affecting the abilities we once knew as innately human perception, rationalisation, and intellect, creating powerful technologies and intelligent systems that allow for better decision-making in dynamic and complex environments. Rapid developments in sensing and autonomous technologies are transforming industries (Lee and Qiufan, 2024). Moving forward, there will be extraordinary research opportunities to enhance and re-envision these technologies and systems to harness their potential and help make real-time decisions based on combinations of quantitative and qualitative data.

Bibliography

Bartneck, C. *et al.* (2024) *Human-Robot Interaction: An Introduction.* Cambridge: Cambridge University Press.

Dhivya, M. *et al.* (2021) *Artificial Intelligence (AI): Recent Trends and Applications*. Boca Raton: CRC Press.

Gamito, M. and Eber, M. (2029) *Algorithmic Governance and Governance of Algorithms: Legal and Ethical Challenges*. Wiesbaden: Springer.

Helmold, M. (2023) *Wettbewerbsvorteile entlang der Supply Chain sichern: Best-Practice-Beispiele in Beschaffung, Produktion, Marketing und anderen Funktionen der betriebswirtschaftlichen Wertschöpfungskette*. Wiesbaden: Springer.

Helmold, M. (2024) *Erfolgreiche Transformation zum digitalen Champion: Wettbewerbsvorteile durch Digitalisierung und Künstliche Intelligenz*. Wiesbaden: Springer.

Hohoff, C. (2024) *Kompetenzen für die Arbeitswelten der Zukunft: Impulse des European Year of Skills für Wirtschaft, Bildung und Personalwesen*. Wiesbaden: Springer.

Lee, K.F. and Qiufan, C. (2024) *AI 2041: Ten Visions for Our Future*. New York: Crown Publishing.

Mucci, T. and Gather, W. (2025) 'Die Zukunft der KI: Trends, die die nächsten 10 Jahre prägen', *IBM*. Available at: https://www.ibm.com/think/author/tim-mucci (Accessed: 2 May 2025).

OECD (2020) *Künstliche Intelligenz in der Gesellschaft*. Available at: https://www.google.de/books/edition/_/aadHzgEACAAJ?hl=en&sa=X&ved=2ahUKEwj8yObD4IaNAxUCQ_EDHZsEAnQQ8fIDegQIExAM (Accessed: 3 May 2025).

Staff, J. and Groß, M. (2024) *KI-Revolution der Arbeitswelt: Perspektiven für Management, Organisation und HR. Auswirkungen, Einfluss, Chancen von Künstlicher Intelligenz auf Berufsbilder und Arbeitsformen*. München: Haufe.

Verma, M. *et al*. (2024) *Recent Trends in Artificial Intelligence Towards a Smart World: Applications in Industries and Sectors*. Singapore: Springer Nature.

Index

I

L

M

N

P

R

S

T

V

www.ingramcontent.com/pod-product-compliance
Lightning Source LLC
LaVergne TN
LVHW011811200926
848484LV00005B/1152
9781800618046